Yamaha TY50, 80, 125 & 175 Owners Workshop Manual

by Jeremy Churchill

Models covered

TY50 P. 49cc. Introduced July 1976, discontinued August 1977
TY50 M. 49cc. Introduced August 1977, discontinued April 1983
TY80. 72cc. Introduced 1974, discontinued 1984
TY125. 123cc. Introduced April 1982, discontinued 1984
TY175. 171cc. Introduced 1975, discontinued 1984

This manual will also be of use to owners of the Yamaha based trials specials

ISBN 978 0 85696 464 0

© Haynes Publishing Group 1990

ABCDE
FGHIJ
KLMNO

2

Printed in Malaysia *(464 – 5P2)*

Haynes Publishing Group
Sparkford Nr Yeovil
Somerset BA22 7JJ England

Haynes Publications, Inc
859 Lawrence Drive
Newbury Park
California 91320 USA

*Printed using NORBRITE BOOK 48.8gsm (CODE:
40N6533) from NORPAC; procurement system
certified under Sustainable Forestry Initiative
standard. Paper produced is certified to the
SFI Certified Fiber Sourcing Standard (CERT -
0094271)*

British Library Cataloguing in Publication Data

Churchill, Jeremy
    Yamaha TY series owners workshop manual.–
(Owners workshop manual)
1. Yamaha motorcycle
I. Title
629.28'775   TL448.Y3
ISBN 0–85696–464–6

Acknowledgements

Special thanks are due to Colin Appleyard of Colin Appleyard Motorcycles, Keighley, Yorkshire who supplied the machine featured in the photographs throughout this manual, and to Roger Painter of Roger Painter Competition Motorcycles, Warminster, Wiltshire, for the use of his TY175. Thanks are due to Mitsui Machinery Sales (UK) Ltd who supplied the service information and gave permission to reproduce many of the line drawings and particularly to Phil Kibler and Bob Gower of Mitsui's Technical Support Group for their invaluable assistance in finding scarce information.

We would also like to thank the Avon Rubber Company, who kindly supplied information and technical assistance on tyre fitting, NGK Spark Plugs (UK) Ltd for information on spark plug maintenance and electrode conditions, and Renold Ltd for advice on chain care and renewal.

About this manual

The purpose of this manual is to present the owner with a concise and graphic guide which will enable him to tackle any operation from basic routine maintenance to a major overhaul. It has been assumed that any work would be undertaken without the luxury of a well-equipped workshop and a range of manufacturer's service tools.

To this end, the machine featured in the manual was stripped and rebuilt in our own workshop, by a team comprising a mechanic, a photographer and the author. The resulting photographic sequence depicts events as they took place, the hands shown being those of the author and the mechanic.

The use of specialised, and expensive, service tools was avoided unless their use was considered to be essential due to risk of breakage or injury. There is usually some way of improvising a method of removing a stubborn component, providing that a suitable degree of care is exercised.

The author learnt his motorcycle mechanics over a number of years, faced with the same difficulties and using similar facilities to those encountered by most owners. It is hoped that this practical experience can be passed on through the pages of this manual.

Where possible, a well-used example of the machine is chosen for the workshop project, as this highlights any areas which might be particularly prone to giving rise to problems. In this way, any such difficulties are encountered and resolved before the text is written, and the techniques used to deal with them can be incorporated in the relevant section. Armed with a working knowledge of the machine, the author undertakes a considerable amount of research in order that the maximum amount of data can be included in the manual.

A comprehensive section, preceding the main part of the manual, describes procedures for carrying out the routine maintenance of the machine at intervals of time and mileage. This section is included particularly for those owners who wish to ensure the efficient day-to-day running of their motorcycle, but who choose not to undertake overhaul or renovation work.

Each Chapter is divided into numbered sections. Within these sections are numbered paragraphs. Cross reference throughout the manual is quite straightforward and logical. When reference is made 'See Section 6.10' it means Section 6, paragraph 10 in the same Chapter. If another Chapter were intended, the reference would read, for example, 'See Chapter 2, Section 6.10'. All the photographs are captioned with a section/paragraph number to which they refer and are relevant to the Chapter text adjacent.

Figures (usually line illustrations) appear in a logical but numerical order, within a given Chapter. Fig. 1.1 therefore refers to the first figure in Chapter 1.

Left-hand and right-hand descriptions of the machines and their components refer to the left and right of a given machine when the rider is seated normally.

Motorcycle manufacturers continually make changes to specifications and recommendations, and these, when notified, are incorporated into our manuals at the earliest opportunity.

Whilst every care is taken to ensure that the information in this manual is correct no liability can be accepted by the author or publishers for loss, damage or injury caused by any errors in or omissions from the information given.

Contents

Left-hand view of the TY175

Right-hand view of the TY175

Engine/gearbox unit

Introduction to the Yamaha TY models

The 49cc models covered in this Manual were introduced in July 1976 in the form of the TY50 P. The engine/gearbox unit was identical in design to other models in Yamaha's range, notably the TY80, and the machine was given its own individual identity by the use of trials-orientated styling in the cycle parts. To comply with legislation in force in the UK at the time the machine was fitted with pedals instead of footrests. Although the softly-tuned engine placed the machine at a disadvantage initially when compared with its European rivals, the situation changed when, in 1977, a moped was re-defined as a lightweight motorcycle of no more than 50cc engine capacity and a designed maximum speed of 30 mph. While other manufacturers had to spend time and money restricting their existing models' power outputs or developing new models, Yamaha had merely to substitute footrests for the pedals. The new version, the TY50 M, was introduced in August 1977 and was sold until it was discontinued in April 1983. The only major modifications were to introduce larger diameter wheels in 1978, this necessitating a change in gearing, and to reduce the number of plates in the clutch.

The TY80 and TY175 models were introduced at a time when trials riding was increasing in popularity, and, in conjunction with the 250 model developed by Mick Andrews, were intended to gain Yamaha as large a share as possible of the new market. Although trials machines have never sold well enough to encourage the big four Japanese factories into the level of competition evident in motocross, for example, these two Yamaha models were bought by relatively large numbers of riders seeking an alternative to the more expensive European machinery. Although not as competitive in standard form as the more specialized European products, the TY80 and 175 are perfectly adequate for the rider of normal ability and have become extremely popular due to their cheapness and their ease of maintenance. For those seeking to improve their machines for trials use there are many aftermarket products available from specialist trials dealers, ranging from alloy handlebars to complete frame kits. In spite of the rapidly increasing sophistication of current trials models, a glance through the results lists in the motorcycle press will show a high proportion of Yamaha models or Yamaha-based specials among the winners.

The 80 and 175 models continued virtually unchanged for nearly ten years of production; a feat almost unheard of in Japanese motorcycle production and proof positive of Yamaha's satisfaction with them. It came as something of a surprise, therefore, when it was announced in November 1984 that both were to be discontinued.

This leaves only the TY125 in production; introduced in April 1982 it is virtually identical to the TY175 model except for the obvious difference in engine capacity. It has been fitted with a full battery-powered electrical system which includes turn signals, in an attempt to attract the learner riders who are currently restricted to machines of 125cc. Although this road-going equipment carries a weight penalty of over 20 lbs, it can easily be removed for serious trials use, whilst the same large range of aftermarket products available for the 175 can also be used on this model.

Model dimensions and weights

	TY50 P, early TY50 M	Late TY50 M	TY80	TY125	TY175
Overall length	1800 mm (70.9 in), 1815 mm (71.5 in) TY50 M	1860 mm (73.2 in)	1560 mm (61.4 in)	1995 mm (78.5 in)	1955 mm (77.0 in)
Overall width	775 mm (30.5 in)	775 mm (30.5 in)	690 mm (27.2 in)	820 mm (32.3 in)	835 mm (32.9 in)
Overall height	975 mm (38.3 in)	N/Av	890 mm (35.0 in)	1120 mm (44.1 in)	1100 mm (43.3 in)
Wheelbase	1190 mm (46.9 in), 1210 mm (47.6 in) TY50 M	1210 mm (47.6 in)	1025 mm (40.4 in)	1265 mm (49.8 in)	1265 mm (49.8 in)
Seat height	720 mm (28.4 in)	765 mm (30.1 in)	600 mm (23.6 in)	775 mm (30.5 in)	750 mm (29.5 in)
Ground clearance	200 mm (7.9 in)	225 mm (8.9 in)	220 mm (8.7 in)	295 mm (11.6 in)	295 mm (11.6 in)
Weight	73 kg (161 lb)	72 kg (159 lb)	54 kg (119 lb)	93 kg (205 lb)	81 kg (179 lb)

Ordering spare parts

Before attempting any overhaul or maintenance work it is important to ensure that any parts likely to be required are to hand. Many of the more common parts such as gaskets and seals will be available off the shelf from the local Yamaha dealer, but often it will prove necessary to order more specialised parts well in advance. It is worthwhile running through the operation to be undertaken, referring to the appropriate Chapter and Section of this book, so that a note can be made of the items most likely to be required. In some instances it will of course be necessary to dismantle the assembly in question so that the various components can be examined and measured for wear and in these instances, it must be remembered that the machine may have to be left dismantled while the replacement parts are obtained.

It is advisable to purchase almost all new parts from an official Yamaha dealer. Almost any motorcycle dealer should be able to obtain the parts in time, but this may take longer than it would through the official factory spares arrangement. It is quite in order to purchase expendable items such as spark plugs, bulbs, tyres, oil and grease from the nearest convenient source.

Owners should be very wary of some of the pattern parts that might be offered at a lower price than the Yamaha originals. Whilst in most cases these will be of an adequate standard, some of the more important parts have been known to fail suddenly and cause extensive damage in the process. A particular danger in recent years is the growing number of counterfeit parts from Taiwan. These include items such as oil filters and brake pads and are often sold in packaging which is almost indistinguishable from the manufacturer's own. Again, these are often quite serviceable parts, but can sometimes be dangerously inadequate in materials or construction. Apart from rendering the manufacturer's warranty invalid, use of sub-standard parts may put the life of the rider (or the machine) at risk. In short, where there are any doubts on safety grounds, purchase parts **only** from a reputable Yamaha dealer. The extra cost involved pays for a high standard of quality and the parts will be guaranteed to work effectively.

Most machines are subject to continuous detail modifications throughout their production run in addition to annual model changes. In most cases these changes will be known to the dealer but not to the general public, so it is essential to quote the engine and frame numbers in full when ordering parts. The engine number is embossed in a rectangular section of the crankcase just below the carburettor, and the frame number is stamped on the right-hand side of the steering head.

Location of engine number

Location of frame number

Safety first!

Professional motor mechanics are trained in safe working procedures. However enthusiastic you may be about getting on with the job in hand, do take the time to ensure that your safety is not put at risk. A moment's lack of attention can result in an accident, as can failure to observe certain elementary precautions.

There will always be new ways of having accidents, and the following points do not pretend to be a comprehensive list of all dangers; they are intended rather to make you aware of the risks and to encourage a safety-conscious approach to all work you carry out on your vehicle.

Essential DOs and DON'Ts

DON'T start the engine without first ascertaining that the transmission is in neutral.

DON'T suddenly remove the filler cap from a hot cooling system – cover it with a cloth and release the pressure gradually first, or you may get scalded by escaping coolant.

DON'T attempt to drain oil until you are sure it has cooled sufficiently to avoid scalding you.

DON'T grasp any part of the engine, exhaust or silencer without first ascertaining that it is sufficiently cool to avoid burning you.

DON'T allow brake fluid or antifreeze to contact the machine's paintwork or plastic components.

DON'T syphon toxic liquids such as fuel, brake fluid or antifreeze by mouth, or allow them to remain on your skin.

DON'T inhale dust – it may be injurious to health (see *Asbestos* heading).

DON'T allow any spilt oil or grease to remain on the floor – wipe it up straight away, before someone slips on it.

DON'T use ill-fitting spanners or other tools which may slip and cause injury.

DON'T attempt to lift a heavy component which may be beyond your capability – get assistance.

DON'T rush to finish a job, or take unverified short cuts.

DON'T allow children or animals in or around an unattended vehicle.

DON'T inflate a tyre to a pressure above the recommended maximum. Apart from overstressing the carcase and wheel rim, in extreme cases the tyre may blow off forcibly.

DO ensure that the machine is supported securely at all times. This is especially important when the machine is blocked up to aid wheel or fork removal.

DO take care when attempting to slacken a stubborn nut or bolt. It is generally better to pull on a spanner, rather than push, so that if slippage occurs you fall away from the machine rather than on to it.

DO wear eye protection when using power tools such as drill, sander, bench grinder etc.

DO use a barrier cream on your hands prior to undertaking dirty jobs – it will protect your skin from infection as well as making the dirt easier to remove afterwards; but make sure your hands aren't left slippery. Note that long-term contact with used engine oil can be a health hazard.

DO keep loose clothing (cuffs, tie etc) and long hair well out of the way of moving mechanical parts.

DO remove rings, wristwatch etc, before working on the vehicle – especially the electrical system.

DO keep your work area tidy – it is only too easy to fall over articles left lying around.

DO exercise caution when compressing springs for removal or installation. Ensure that the tension is applied and released in a controlled manner, using suitable tools which preclude the possibility of the spring escaping violently.

DO ensure that any lifting tackle used has a safe working load rating adequate for the job.

DO get someone to check periodically that all is well, when working alone on the vehicle.

DO carry out work in a logical sequence and check that everything is correctly assembled and tightened afterwards.

DO remember that your vehicle's safety affects that of yourself and others. If in doubt on any point, get specialist advice.

IF, in spite of following these precautions, you are unfortunate enough to injure yourself, seek medical attention as soon as possible.

Asbestos

Certain friction, insulating, sealing, and other products – such as brake linings, clutch linings, gaskets, etc – contain asbestos. *Extreme care must be taken to avoid inhalation of dust from such products since it is hazardous to health.* If in doubt, assume that they *do* contain asbestos.

Fire

Remember at all times that petrol (gasoline) is highly flammable. Never smoke, or have any kind of naked flame around, when working on the vehicle. But the risk does not end there – a spark caused by an electrical short-circuit, by two metal surfaces contacting each other, by careless use of tools, or even by static electricity built up in your body under certain conditions, can ignite petrol vapour, which in a confined space is highly explosive.

Always disconnect the battery earth (ground) terminal before working on any part of the fuel or electrical system, and never risk spilling fuel on to a hot engine or exhaust.

It is recommended that a fire extinguisher of a type suitable for fuel and electrical fires is kept handy in the garage or workplace at all times. Never try to extinguish a fuel or electrical fire with water.

Note: *Any reference to a 'torch' appearing in this manual should always be taken to mean a hand-held battery-operated electric lamp or flashlight. It does **not** mean a welding/gas torch or blowlamp.*

Fumes

Certain fumes are highly toxic and can quickly cause unconsciousness and even death if inhaled to any extent. Petrol (gasoline) vapour comes into this category, as do the vapours from certain solvents such as trichloroethylene. Any draining or pouring of such volatile fluids should be done in a well ventilated area.

When using cleaning fluids and solvents, read the instructions carefully. Never use materials from unmarked containers – they may give off poisonous vapours.

Never run the engine of a motor vehicle in an enclosed space such as a garage. Exhaust fumes contain carbon monoxide which is extremely poisonous; if you need to run the engine, always do so in the open air or at least have the rear of the vehicle outside the workplace.

The battery

Never cause a spark, or allow a naked light, near the vehicle's battery. It will normally be giving off a certain amount of hydrogen gas, which is highly explosive.

Always disconnect the battery earth (ground) terminal before working on the fuel or electrical systems.

If possible, loosen the filler plugs or cover when charging the battery from an external source. Do not charge at an excessive rate or the battery may burst.

Take care when topping up and when carrying the battery. The acid electrolyte, even when diluted, is very corrosive and should not be allowed to contact the eyes or skin.

If you ever need to prepare electrolyte yourself, always add the acid slowly to the water, and never the other way round. Protect against splashes by wearing rubber gloves and goggles.

Mains electricity and electrical equipment

When using an electric power tool, inspection light etc, always ensure that the appliance is correctly connected to its plug and that, where necessary, it is properly earthed (grounded). Do not use such appliances in damp conditions and, again, beware of creating a spark or applying excessive heat in the vicinity of fuel or fuel vapour. Also ensure that the appliances meet the relevant national safety standards.

Ignition HT voltage

A severe electric shock can result from touching certain parts of the ignition system, such as the HT leads, when the engine is running or being cranked, particularly if components are damp or the insulation is defective. Where an electronic ignition system is fitted, the HT voltage is much higher and could prove fatal.

Tools and working facilities

The first priority when undertaking maintenance or repair work of any sort on a motorcycle is to have a clean, dry, well-lit working area. Work carried out in peace and quiet in the well-ordered atmosphere of a good workshop will give more satisfaction and much better results than can usually be achieved in poor working conditions. A good workshop must have a clean flat workbench or a solidly constructed table of convenient working height. The workbench or table should be equipped with a vice which has a jaw opening of at least 4 in (100 mm). A set of jaw covers should be made from soft metal such as aluminium alloy or copper, or from wood. These covers will minimise the marking or damaging of soft or delicate components which may be clamped in the vice. Some clean, dry, storage space will be required for tools, lubricants and dismantled components. It will be necessary during a major overhaul to lay out engine/gearbox components for examination and to keep them where they will remain undisturbed for as long as is necessary. To this end it is recommended that a supply of metal or plastic containers of suitable size is collected. A supply of clean, lint-free, rags for cleaning purposes and some newspapers, other rags, or paper towels for mopping up spillages should also be kept. If working on a hard concrete floor note that both the floor and one's knees can be protected from oil spillages and wear by cutting open a large cardboard box and spreading it flat on the floor under the machine or workbench. This also helps to provide some warmth in winter and to prevent the loss of nuts, washers, and other tiny components which have a tendency to disappear when dropped on anything other than a perfectly clean, flat, surface.

Unfortunately, such working conditions are not always available to the home mechanic. When working in poor conditions it is essential to take extra time and care to ensure that the components being worked on are kept scrupulously clean and to ensure that no components or tools are lost or damaged.

A selection of good tools is a fundamental requirement for anyone contemplating the maintenance and repair of a motor vehicle. For the owner who does not possess any, their purchase will prove a considerable expense, offsetting some of the savings made by doing-it-yourself. However, provided that the tools purchased meet the relevant national safety standards and are of good quality, they will last for many years and prove an extremely worthwhile investment.

To help the average owner to decide which tools are needed to carry out the various tasks detailed in this manual, we have compiled three lists of tools under the following headings: *Maintenance and minor repair, Repair and overhaul,* and *Specialized.* The newcomer to practical mechanics should start off with the simpler jobs around the vehicle. Then, as his confidence and experience grow, he can undertake more difficult tasks, buying extra tools as and when they are needed. In this way, a *Maintenance and minor repair* tool kit can be built-up into a *Repair and overhaul* tool kit over a considerable period of time without any major cash outlays. The experienced home mechanic will have a tool kit good enough for most repair and overhaul procedures and will add tools from the specialized category when he feels the expense is justified by the amount of use these tools will be put to.

It is obviously not possible to cover the subject of tools fully here. For those who wish to learn more about tools and their use there is a book entitled *Motorcycle Workshop Practice Manual* (Bk no. 1454) available from the publishers of this manual.

As a general rule, it is better to buy the more expensive, good quality tools. Given reasonable use, such tools will last for a very long time, whereas the cheaper, poor quality, item will wear out faster and need to be renewed more often, thus nullifying the original saving. There is also the risk of a poor quality tool breaking while in use, causing personal injury or expensive damage to the component being worked on.

For practically all tools, a tool factor is the best source since he will have a very comprehensive range compared with the average garage or accessory shop. Having said that, accessory shops often offer excellent quality tools at discount prices, so it pays to shop around. There are plenty of tools around at reasonable prices, but always aim to purchase items which meet the relevant national safety standards. If in doubt, seek the advice of the shop proprietor or manager before making a purchase.

The basis of any toolkit is a set of spanners. While open-ended spanners with their slim jaws, are useful for working on awkwardly-positioned nuts, ring spanners have advantages in that they grip the nut far more positively. There is less risk of the spanner slipping off the nut and damaging it, for this reason alone ring spanners are to be preferred. Ideally, the home mechanic should acquire a set of each, but if expense rules this out a set of combination spanners (open-ended at one end and with a ring of the same size at the other) will provide a good compromise. Another item which is so useful it should be considered an essential requirement for any home mechanic is a set of socket spanners. These are available in a variety of drive sizes. It is recommended that the ½-inch drive type is purchased to begin with as although bulkier and more expensive than the ⅜-inch type, the larger size is far more common and will accept a greater variety of torque wrenches, extension pieces and socket sizes. The socket set should comprise sockets of sizes between 8 and 24 mm, a reversible ratchet drive, an extension bar of about 10 inches in length, a spark plug socket with a rubber insert, and a universal joint. Other attachments can be added to the set at a later date.

Maintenance and minor repair tool kit

Set of spanners 8 – 24 mm
Set of sockets and attachments
Spark plug spanner with rubber insert – 10, 12, or 14 mm as appropriate
Adjustable spanner
C-spanner/pin spanner
Torque wrench (same size drive as sockets)
Set of screwdrivers (flat blade)
Set of screwdrivers (cross-head)
Set of Allen keys 4 – 10 mm
Impact screwdriver and bits
Ball pein hammer – 2 lb
Hacksaw (junior)
Self-locking pliers – Mole grips or vice grips
Pliers – combination
Pliers – needle nose
Wire brush (small)
Soft-bristled brush
Tyre pump
Tyre pressure gauge
Tyre tread depth gauge
Oil can
Fine emery cloth
Funnel (medium size)
Drip tray
Grease gun
Set of feeler gauges
Brake bleeding kit
Strobe timing light
Continuity tester (dry battery and bulb)
Soldering iron and solder
Wire stripper or craft knife
PVC insulating tape
Assortment of split pins, nuts, bolts, and washers

Repair and overhaul toolkit

The tools in this list are virtually essential for anyone undertaking major repairs to a motorcycle and are additional to the tools listed above. Concerning Torx driver bits, Torx screws are encountered on some of the more modern machines where their use is restricted to fastening certain components inside the engine/gearbox unit. It is therefore recommended that if Torx bits cannot be borrowed from a local dealer, they are purchased individually as the need arises. They are not in regular use in the motor trade and will therefore only be available in specialist tool shops.

Plastic or rubber soft-faced mallet
Torx driver bits
Pliers – electrician's side cutters
Circlip pliers – internal (straight or right-angled tips are available)
Circlip pliers – external
Cold chisel
Centre punch
Pin punch
Scriber
Scraper (made from soft metal such as aluminium or copper)
Soft metal drift
Steel rule/straight edge
Assortment of files
Electric drill and bits

Wire brush (large)
Soft wire brush (similar to those used for cleaning suede shoes)
Sheet of plate glass
Hacksaw (large)
Valve grinding tool
Valve grinding compound (coarse and fine)
Stud extractor set (E-Z out)

Specialized tools

This is not a list of the tools made by the machine's manufacturer to carry out a specific task on a limited range of models. Occasional references are made to such tools in the text of this manual and, in general, an alternative method of carrying out the task without the manufacturer's tool is given where possible. The tools mentioned in this list are those which are not used regularly and are expensive to buy in view of their infrequent use. Where this is the case it may be possible to hire or borrow the tools against a deposit from a local dealer or tool hire shop. An alternative is for a group of friends or a motorcycle club to join in the purchase.

Valve spring compressor
Piston ring compressor
Universal bearing puller
Cylinder bore honing attachment (for electric drill)
Micrometer set
Vernier calipers
Dial gauge set
Cylinder compression gauge
Vacuum gauge set
Multimeter
Dwell meter/tachometer

Care and maintenance of tools

Whatever the quality of the tools purchased, they will last much longer if cared for. This means in practice ensuring that a tool is used for its intended purpose; for example screwdrivers should not be used as a substitute for a centre punch, or as chisels. Always remove dirt or grease and any metal particles but remember that a light film of oil will prevent rusting if the tools are infrequently used. The common tools can be kept together in a large box or tray but the more delicate, and more expensive, items should be stored separately where they cannot be damaged. When a tool is damaged or worn out, be sure to renew it immediately. It is false economy to continue to use a worn spanner or screwdriver which may slip and cause expensive damage to the component being worked on.

Fastening systems

Fasteners, basically, are nuts, bolts and screws used to hold two or more parts together. There are a few things to keep in mind when working with fasteners. Almost all of them use a locking device of some type: either a lock washer, lock nut, locking tab or thread adhesive. All threaded fasteners should be clean, straight, have undamaged threads and undamaged corners on the hexagon head where the spanner fits. Develop the habit of replacing all damaged nuts and bolts with new ones.

Rusted nuts and bolts should be treated with a rust penetrating fluid to ease removal and prevent breakage. After applying the rust penetrant, let it 'work' for a few minutes before trying to loosen the nut or bolt. Badly rusted fasteners may have to be chiseled off or removed with a special nut breaker, available at tool shops.

Flat washers and lock washers, when removed from an assembly should always be replaced exactly as removed. Replace any damaged washers with new ones. Always use a flat washer between a lock washer and any soft metal surface (such as aluminium), thin sheet metal or plastic. Special lock nuts can only be used once or twice before they lose their locking ability and must be renewed.

If a bolt or stud breaks off in an assembly, it can be drilled out and removed with a special tool called an E-Z out. Most dealer service departments and motorcycle repair shops can perform this task, as well as others (such as the repair of threaded holes that have been stripped out).

Spanner size comparison

Jaw gap (in)	Spanner size	Jaw gap (in)	Spanner size
0.250	$\frac{1}{4}$ in AF	0.945	24 mm
0.276	7 mm	1.000	1 in AF
0.313	$\frac{5}{16}$ in AF	1.010	$\frac{9}{16}$ in Whitworth; $\frac{5}{8}$ in BSF
0.315	8 mm	1.024	26 mm
0.344	$\frac{11}{32}$ in AF; $\frac{1}{8}$ in Whitworth	1.063	$1\frac{1}{16}$ in AF; 27 mm
0.354	9 mm	1.100	$\frac{5}{8}$ in Whitworth; $\frac{11}{16}$ in BSF
0.375	$\frac{3}{8}$ in AF	1.125	$1\frac{1}{8}$ in AF
0.394	10 mm	1.181	30 mm
0.433	11 mm	1.200	$\frac{11}{16}$ in Whitworth; $\frac{3}{4}$ in BSF
0.438	$\frac{7}{16}$ in AF	1.250	$1\frac{1}{4}$ in AF
0.445	$\frac{3}{16}$ in Whitworth; $\frac{1}{4}$ in BSF	1.260	32 mm
0.472	12 mm	1.300	$\frac{3}{4}$ in Whitworth; $\frac{7}{8}$ in BSF
0.500	$\frac{1}{2}$ in AF	1.313	$1\frac{5}{16}$ in AF
0.512	13 mm	1.390	$\frac{13}{16}$ in Whitworth; $\frac{15}{16}$ in BSF
0.525	$\frac{1}{4}$ in Whitworth; $\frac{5}{16}$ in BSF	1.417	36 mm
0.551	14 mm	1.438	$1\frac{7}{16}$ in AF
0.563	$\frac{9}{16}$ in AF	1.480	$\frac{7}{8}$ in Whitworth; 1 in BSF
0.591	15 mm	1.500	$1\frac{1}{2}$ in AF
0.600	$\frac{5}{16}$ in Whitworth; $\frac{3}{8}$ in BSF	1.575	40 mm; $\frac{15}{16}$ in Whitworth
0.625	$\frac{5}{8}$ in AF	1.614	41 mm
0.630	16 mm	1.625	$1\frac{5}{8}$ in AF
0.669	17 mm	1.670	1 in Whitworth; $1\frac{1}{8}$ in BSF
0.686	$\frac{11}{16}$ in AF	1.688	$1\frac{11}{16}$ in AF
0.709	18 mm	1.811	46 mm
0.710	$\frac{3}{8}$ in Whitworth; $\frac{7}{16}$ in BSF	1.813	$1\frac{13}{16}$ in AF
0.748	19 mm	1.860	$1\frac{1}{8}$ in Whitworth; $1\frac{1}{4}$ in BSF
0.750	$\frac{3}{4}$ in AF	1.875	$1\frac{7}{8}$ in AF
0.813	$\frac{13}{16}$ in AF	1.969	50 mm
0.820	$\frac{7}{16}$ in Whitworth; $\frac{1}{2}$ in BSF	2.000	2 in AF
0.866	22 mm	2.050	$1\frac{1}{4}$ in Whitworth; $1\frac{3}{8}$ in BSF
0.875	$\frac{7}{8}$ in AF	2.165	55 mm
0.920	$\frac{1}{2}$ in Whitworth; $\frac{9}{16}$ in BSF	2.362	60 mm
0.938	$\frac{15}{16}$ in AF		

Standard torque settings

Specific torque settings will be found at the end of the specifications section of each chapter. Where no figure is given, bolts should be secured according to the table below.

Fastener type (thread diameter)	kgf m	lbf ft
5mm bolt or nut	0.45 – 0.6	3.5 – 4.5
6 mm bolt or nut	0.8 – 1.2	6 – 9
8 mm bolt or nut	1.8 – 2.5	13 – 18
10 mm bolt or nut	3.0 – 4.0	22 – 29
12 mm bolt or nut	5.0 – 6.0	36 – 43
5 mm screw	0.35 – 0.5	2.5 – 3.6
6 mm screw	0.7 – 1.1	5 – 8
6 mm flange bolt	1.0 – 1.4	7 – 10
8 mm flange bolt	2.4 – 3.0	17 – 22
10 mm flange bolt	3.0 – 4.0	22 – 29

Fault diagnosis

Contents

1 Introduction

This Section provides an easy reference-guide to the more common ailments that are likely to afflict your machine. Obviously, the opportunities are almost limitless for faults to occur as a result of obscure failures, and to try and cover all eventualities would require a book. Indeed, a number have been written on the subject.

Successful fault diagnosis is not a mysterious 'black art' but the application of a bit of knowledge combined with a systematic and logical approach to the problem. Approach any fault diagnosis by first accurately identifying the symptom and then checking through the list of possibile causes, starting with the simplest or most obvious and progressing in stages to the most complex. Take nothing for granted, but above all apply liberal quantities of common sense.

The main symptom of a fault is given in the text as a major heading below which are listed, as Sections headings, the various systems or areas which may contain the fault. Details of each possible cause for a fault and the remedial action to be taken are given, in brief, in the paragraphs below each Section heading. Further information should be sought in the relevant Chapter.

Engine does not start when turned over

2 No fuel flow to carburettor

● Fuel tank empty or level too low. Check that the tap is turned to 'On' or 'Reserve' position as required. If in doubt, prise off the fuel feed pipe at the carburettor end and check that fuel runs from pipe when the tap is turned on.

● Tank filler cap vent obstructed. This can prevent fuel from flowing into the carburettor float bowl bcause air cannot enter the fuel tank to replace it. The problem is more likely to appear when the machine is being ridden. Check by listening close to the filler cap and releasing it. A hissing noise indicates that a blockage is present. Remove the cap and clear the vent hole with wire or by using an air line from the inside of the cap.

● Fuel tap or filter blocked. Blockage may be due to accumulation of rust or paint flakes from the tank's inner surface or of foreign matter from contaminated fuel. Remove the tap and clean it and the filter. Look also for water droplets in the fuel.

● Fuel line blocked. Blockage of the fuel line is more likely to result from a kink in the line rather than the accumulation of debris.

3 Fuel not reaching cylinder

● Float chamber not filling. Caused by float needle or floats sticking in up position. This may occur after the machine has been left standing for an extended length of time allowing the fuel to evaporate. When this occurs a gummy residue is often left which hardens to a varnish-like substance. This condition may be worsened by corrosion and crystalline deposits produced prior to the total evaporation of contaminated fuel. Sticking of the float needle may also be caused by wear. In any case removal of the float chamber will be necessary for inspection and cleaning.

● Blockage in starting circuit, slow running circuit or jets. Blockage of these items may be attributable to debris from the fuel tank by-passing the filter system or to gumming up as described in paragraph 1. Water droplets in the fuel will also block jets and passages. The carburettor should be dismantled for cleaning.

● Fuel level too low. The fuel level in the float chamber is controlled by float height. The fuel level may increase with wear or damage but will never reduce, thus a low fuel level is an inherent rather than developing condition. Check the float height, renewing the float or needle if required.

● Oil blockage in fuel system or carburettor (petroil lubricated engines only). May arise when the machine has been parked for long periods and the residual petrol has evaporated. To rectify, dismantle and clean the carburettor and tap, flush the tank and fill with fresh petroil mixed in the correct proportions. This problem can be avoided by running the float bowl dry before the machine is stored for long periods. Do not attempt to use fuel which has become stale.

4 Engine flooding

● Float valve needle worn or stuck open. A piece of rust or other debris can prevent correct seating of the needle against the valve seat thereby permitting an uncontrolled flow of fuel. Similarly, a worn needle or needle seat will prevent valve closure. Dismantle the carburettor float bowl for cleaning and, if necessary, renewal of the worn components.

● Fuel level too high. The fuel level is controlled by the float height which may increase due to wear of the float needle, pivot pin or operating tang. Check the float height, and make any necessary adjustments. A leaking float will cause an increase in fuel level, and thus should be renewed.

● Cold starting mechanism. Check the choke (starter mechanism) for correct operation. If the mechanism jams in the 'On' position subsequent starting of a hot engine will be difficult.

● Blocked air filter. A badly restricted air filter will cause flooding. Check the filter and clean or renew as required. A collapsed inlet hose will have a similar effect. Check that the air filter inlet has not become blocked by a rag or similar item.

5 No spark at plug

● Ignition switch not on.

● Engine stop switch off.

● Spark plug dirty, oiled or 'whiskered'. Because the induction mixture of a two-stroke engine is inclined to be of a rather oily nature it is comparatively easy to foul the plug electrodes, especially where there have been repeated attempts to start the engine. A machine used for short journeys will be more prone to fouling because the engine may never reach full operating temperature, and the deposits will not burn off. On rare occasions a change of plug grade may be required but the advice of a dealer should be sought before making such a change. 'Whiskering' is a comparatively rare occurrence on modern machines but may be encountered where pre-mixed petrol and oil (petroil) lubrication is employed. An electrode deposit in the form of a barely visible filament across the plug electrodes can short circuit the plug and prevent ifs sparking. On all two-stroke machines it is a sound precaution to carry a new spare spark plug for substitution in the event of fouling problems.

● Spark plug failure. Clean the spark plug thoroughly and reset the electrode gap. Refer to the spark plug section and the colour condition guide in Routine Maintenance. If the spark plug shorts internally or has sustained visible damage to the electrodes, core or ceramic insulator it should be renewed. On rare occasions a plug that appears to spark vigorously will fail to do so when refitted to the engine and subjected to the compression pressure in the cylinder.

● Spark plug cap or high tension (HT) lead faulty. Check condition and security. Replace if deterioration is evident. Most spark plug caps have an internal resistor designed to inhibit electrical interference with radio and television sets. On rare occasions the resistor may break down, thus preventing sparking. If this is suspected, fit a new cap as a precaution.

● Spark plug cap loose. Check that the spark plug cap fits securely over the plug and, where fitted, the screwed terminal on the plug end is secure.

● Shorting due to moisture. Certain parts of the ignition system are susceptible to shorting when the machine is ridden or parked in wet weather. Check particularly the area from the spark plug cap back to the ignition coil. A water dispersant spray may be used to dry out waterlogged components. Recurrence of the problem can be prevented by using an ignition sealant spray after drying out and cleaning.

● Ignition or stop switch shorted. May be caused by water corrosion or wear. Water dispersant and contact cleaning sprays may be used. If this fails to overcome the problem dismantling and visual inspection of the switches will be required.

● Shorting or open circuit in wiring. Failure in any wire connecting any of the ignition components will cause ignition malfunction. Check also that all connections are clean, dry and tight.

● Ignition coil failure. Check the coil, referring to Chapter 3.

● Capacitor (condenser) failure. The capacitor may be checked most easily by substitution with a replacement item. Blackened contact breaker points indicate capacitor malfunction but this may not always occur.

● Contact breaker points pitted, burned or closed up. Check the contact breaker points, referring to Routine Maintenance. Check also that the low tension leads at the contact breaker are secure and not shorting out.

6 Weak spark at plug

● Feeble sparking at the plug may be caused by any of the faults mentioned in the preceding Section other than those items in the first three paragraphs. Check first the contact breaker assembly and the spark plug, these being the most likely culprits.

7 Compression low

● Spark plug loose. This will be self-evident on inspection, and may be accompanied by a hissing noise when the engine is turned over. Remove the plug and check that the threads in the cylinder head are not damaged. Check also that the plug sealing washer is in good condition.

● Cylinder head gasket leaking. This condition is often accompanied by a high pitched squeak from around the cylinder head and oil loss, and may be caused by insufficiently tightened cylinder head fasteners, a warped cylinder head or mechanical failure of the gasket material. Re-torqueing the fasteners to the correct specification may seal the leak in some instances but if damage has occurred this course of action will provide, at best, only a temporary cure.

● Low crankcase compression. This can be caused by worn main bearings and seals and will upset the incoming fuel/air mixture. A good seal in these areas is essential on any two-stroke engine.

● Piston rings sticking or broken. Sticking of the piston rings may be caused by seizure due to lack of lubrication or overheating as a result of poor carburation or incorrect fuel type. Gumming of the rings may result from lack of use, or carbon deposits in the ring grooves. Broken rings result from over-revving, over-heating or general wear. In either case a top-end overhaul will be required.

Engine stalls after starting

8 General causes

● Improper cold start mechanism operation. Check that the operating controls function smoothly. A cold engine may not require application of an enriched mixture to start initially but may baulk without choke once firing. Likewise a hot engine may start with an enriched mixture but will stop almost immediately if the choke is inadvertently in operation.

● Ignition malfunction. See Section 9. Weak spark at plug.

● Carburettor incorrectly adjusted. Maladjustment of the mixture strength or idle speed may cause the engine to stop immediately after starting. See Chapter 2.

● Fuel contamination. Check for filter blockage by debris or water which reduces, but does not completely stop, fuel flow, or blockage of the slow speed circuit in the carburettor by the same agents. If water is present it can often be seen as droplets in the bottom of the float bowl. Clean the filter and, where water is in evidence, drain and flush the fuel tank and float bowl.

● Intake air leak. Check for security of the carburettor mounting and hose connections, and for cracks or splits in the hoses. Check also that the carburettor top is secure.

● Air filter blocked or omitted. A blocked filter will cause an over-rich mixture; the omission of a filter will cause an excessively weak mixture. Both conditions will have a detrimental effect on carburation. Clean or renew the filter as necessary.

● Fuel filler cap air vent blocked. Usually caused by dirt or water. Clean the vent orifice.

● Choked exhaust system. Caused by excessive carbon build-up in the system, particularly around the silencer baffles. Refer to Routine Maintenance for further information.

● Excessive carbon build-up in the engine. This can result from failure to decarbonise the engine at the specified interval or through

excessive oil consumption. On pump-fed engines check pump adjustment. On pre-mix (petroil) systems check that oil is mixed in the recommended ratio.

Poor running at idle and low speed

9 Weak spark at plug or erratic firing

● Spark plug fouled, faulty or incorrectly adjusted. See Section 4 or refer to Routine Maintenance.

● Spark plug cap or high tension lead shorting. Check the condition of both these items ensuring that they are in good condition and dry and that the cap is fitted correctly.

● Spark plug type incorrect. Fit plug of correct type and heat range as given in Specifications. In certain conditions a plug of hotter or colder type may be required for normal running.

● Contact breaker points pitted, burned or closed-up. Check the contact breaker assembly, referring to Routine Maintenance.

● Ignition timing incorrect. Check the ignition timing statically and dynamically, ensuring that the advance is functioning correctly.

● Faulty ignition coil. Partial failure of the coil internal insulation will diminish the performance of the coil. No repair is possible, a new component must be fitted.

● Faulty capacitor (condenser). A failure of the capacitor will cause blackening of the contact breaker point faces and will allow excessive sparking at the points. A faulty capacitor may best be checked by substitution of a serviceable replacement item.

● Defective flywheel generator ignition source. Refer to Chapter 3 for further details on test procedures.

10 Fuel/air mixture incorrect

● Intake air leak. Check carburettor mountings and air cleaner hoses for security and signs of splitting. Ensure that carburettor top is tight.

● Mixture strength incorrect. Adjust slow running mixture strength using pilot adjustment screw.

● Pilot jet or slow running circuit blocked. The carburettor should be removed and dismantled for thorough cleaning. Blow through all jets and air passages with compressed air to clear obstructions.

● Air cleaner clogged or omitted. Clean or fit air cleaner element as necessary. Check also that the element and air filter cover are correctly seated.

● Cold start mechanism in operation. Check that the choke has not been left on inadvertently and the operation is correct.

● Fuel level too high or too low. Check the float height, renewing float or needle if required. See Section 3 or 4.

● Fuel tank air vent obstructed. Obstructions usually caused by dirt or water. Clean vent orifice.

11 Compression low

● See Section 7.

Acceleration poor

12 General causes

● All items as for previous Section.

● Choked air filter. Failure to keep the air filter element clean will allow the build-up of dirt with proportional loss of performance. In extreme cases of neglect acceleration will suffer.

● Choked exhaust system. This can result from failure to remove accumulations of carbon from the silencer baffles at the prescribed intervals. The increased back pressure will make the machine noticeably sluggish. Refer to Routine Maintenance for further information on decarbonisation.

● Excessive carbon build-up in the engine. This can result from failure to decarbonise the engine at the specified interval or through excessive oil consumption. On pump-fed engines check pump adjustment. On

pre-mix (petroil) systems check that oil is mixed in the recommended ratio.
● Ignition timing incorrect. Check the contact breaker gap and set within the prescribed range ensuring that the ignition timing is correct. If the contact breaker assembly is worn it may prove impossible to get the gap and timing settings to coincide, necessitating renewal.
● Carburation fault. See Section 10.
● Mechanical resistance. Check that the brakes are not binding. On small machines in particular note that the increased rolling resistance caused by under-inflated tyres may impede acceleration.

Poor running or lack of power at high speeds

13 Weak spark at plug or erratic firing

● All items as for Section 9.
● HT lead insulation failure. Insulation failure of the HT lead and spark plug cap due to old age or damage can cause shorting when the engine is driven hard. This condition may be less noticeable, or not noticeable at all at lower engine speeds.

14 Fuel/air mixture incorrect

● All items as for Section 10, with the exception of items relative exclusively to low speed running.
● Main jet blocked. Debris from contaminated fuel, or from the fuel tank, and water in the fuel can block the main jet. Clean the fuel filter, the float bowl area, and if water is present, flush and refill the fuel tank.
● Main jet is the wrong size. The standard carburettor jetting is for sea level atmospheric pressure. For high altitudes, usually above 5000 ft, a smaller main jet will be required.
● Jet needle and needle jet worn. These can be renewed individually but should be renewed as a pair. Renewal of both items requires partial dismantling of the carburettor.
● Air bleed holes blocked. Dismantle carburettor and use compressed air to blow out all air passages.
● Reduced fuel flow. A reduction in the maximum fuel flow from the fuel tank to the carburettor will cause fuel starvation, proportionate to the engine speed. Check for blockages through debris or a kinked fuel line.

15 Compression low

● See Section 7.

Knocking or pinking

16 General causes

● Carbon build-up in combustion chamber. After high mileages have been covered large accumulations of carbon may occur. This may glow red hot and cause premature ignition of the fuel/air mixture, in advance of normal firing by the spark plug. Cylinder head removal will be required to allow inspection and cleaning.
● Fuel incorrect. A low grade fuel, or one of poor quality may result in compression induced detonation of the fuel resulting in knocking and pinking noises. Old fuel can cause similar problems. A too highly leaded fuel will reduce detonation but will accelerate deposit formation in the combustion chamber and may lead to early pre-ignition as described in item 1.
● Spark plug heat range incorrect. Uncontrolled pre-ignition can result from the use of a spark plug the heat range of which is too hot.
● Weak mixture. Overheating of the engine due to a weak mixture can result in pre-ignition occurring where it would not occur when engine temperature was within normal limits. Maladjustment, blocked jets or passages and air leaks can cause this condition.

Overheating

17 Firing incorrect

● Spark plug fouled, defective or maladjusted. See Section 5.
● Spark plug type incorrect. Refer to the Specifications and ensure that the correct plug type is fitted.
● Incorrect ignition timing. Timing that is far too much advanced or far too much retarded will cause overheating. Check the ignition timing is correct.

18 Fuel/air mixture incorrect

● Slow speed mixture strength incorrect. Adjust pilot air screw.
● Main jet wrong size. The carburettor is jetted for sea level atmospheric conditions. For high altitudes, usually above 5000 ft, a smaller main jet will be required.
● Air filter badly fitted or omitted. Check that the filter element is in place and that it and the air filter box cover are sealing correctly. Any leaks will cause a weak mixture.
● Induction air leaks. Check the security of the carburettor mountings and hose connections, and for cracks and splits in the hoses. Check also that the carburettor top is secure.
● Fuel level too low. See Section 3.
● Fuel tank filler cap air vent obstructed. Clear blockage.

19 Lubrication inadequate

● Petrol/oil mixture incorrect. The proportion of oil mixed with the petrol in the tank is critical if the engine is to perform correctly. Too little oil will leave the reciprocating parts and bearings poorly lubricated and overheating will occur. In extreme case the engine will seize. Conversely, too much oil will effectively displace a similar amount of petrol. Though this does not often cause overheating in practice it is possible that the resultant weak mixture may cause overheating. It will inevitably cause a loss of power and excessive exhaust smoke. This advice applies only if the machine has been converted to petroil lubrication.
● Oil pump settings incorrect. The oil pump settings are of great importance since the quantities of oil being injected are very small. Any variation in oil delivery will have a significant effect on the engine. Refer to Routine Maintenance for further information.
● Oil tank empty or low. This will have disastrous consequences if left unnoticed. Check and replenish tank regularly.
● Transmission oil low or worn out. Check the level regularly and investigate any loss of oil. If the oil level drops with no sign of external leakage it is likely that the crankshaft main bearing oil seals are worn, allowing transmission oil to be drawn into the crankcase during induction.

20 Miscellaneous causes

● Engine fins clogged. A build-up of mud in the cylinder head and cylinder barrel cooling fins will decrease the cooling capabilities of the fins. Clean the fins as required.

Clutch operating problems

21 Clutch slip

● No clutch lever play. Adjust clutch lever end play according to the procedure in Routine Maintenance.
● Friction plates worn or warped. Overhaul clutch assembly, replacing plates out of specification.
● Steel plates worn or warped. Overhaul clutch assembly, replacing plates out of specification.
● Clutch spring broken or worn. Old or heat-damaged (from slipping clutch) springs should be replaced with new ones.
● Clutch release not adjusted properly. See Routine Maintenance.

● Clutch inner cable snagging. Caused by a frayed cable or kinked outer cable. Replace the cable with a new one. Repair of a frayed cable is not advised.
● Clutch release mechanism defective. Replace parts as necessary.
● Clutch hub and outer drum worn. Severe indentation by the clutch plate tangs of the channels in the hub and drum will cause snagging of the plates preventing correct engagement. If this damage occurs, renewal of the worn components is required.
● Lubricant incorrect. Use of a transmission lubricant other than that specified may allow the plates to slip.

22 Clutch drag

● Clutch lever play excessive. Adjust lever at bars or at cable end if necessary.
● Clutch plates warped or damaged. This will cause a drag on the clutch, causing the machine to creep. Overhaul clutch assembly.
● Clutch spring tension uneven. Usually caused by a sagged or broken spring. Check and replace springs.
● Transmission oil deteriorated. Badly contaminated transmission oil and a heavy deposit of oil sludge on the plates will cause plate sticking. The oil recommended for this machine is of the detergent type, therefore it is unlikely that this problem will arise unless regular oil changes are neglected.
● Transmission oil viscosity too high. Drag in the plates will result from the use of an oil with too high a viscosity. In very cold weather clutch drag may occur until the engine has reached operating temperature.
● Clutch hub and outer drum worn. Indentation by the clutch plate tangs of the channels in the hub and drum will prevent easy plate disengagement. If the damage is light the affected areas may be dressed with a fine file. More pronounced damage will necessitate renewal of the components.
● Clutch housing seized to shaft. Lack of lubrication, severe wear or damage can cause the housing to seize to the shaft. Overhaul of the clutch, and perhaps the transmission, may be necessary to repair damage.
● Clutch release mechanism defective. Worn or damaged release mechanism parts can stick and fail to provide leverage.
● Loose clutch centre nut. Causes drum and centre misalignment, putting a drag on the engine. Engagement adjustment continually varies. Overhaul clutch assembly.

Gear selection problems

23 Gear lever does not return

● Weak or broken return spring. Renew the spring.
● Gearchange shaft bent or seized. Distortion of the gearchange shaft often occurs if the machine is dropped heavily on the gear lever. Provided that damage is not severe straightening of the shaft is permissible.

24 Gear selection difficult or impossible

● Clutch not disengaging fully. See Section 22.
● Gearchange shaft bent. This often occurs if the machine is dropped heavily on the gear lever. Straightening of the shaft is permissible if the damage is not too great.
● Gearchange arms or pins worn or damaged. Wear or breakage of any of these items may cause difficulty in selecting one or more gears. Overhaul the selector mechanism.
● Gearchange shaft return spring maladjusted. This is often characterised by difficulties in changing up or down, but rarely in both directions. Adjust the anchor bolt as described in Chapter 1.
● Gearchange drum stopper cam or detent plunger damaged. Failure, rather than wear of these items may jam the drum thereby preventing gearchanging or causing false selection at high speed.
● Selector forks bent or seized. This can be caused by dropping the machine heavily on the gearchange lever or as a result of lack of lubrication. Though rare, bending of a shaft can result from a missed gearchange or false selection at high speed.

● Selector fork end and pin wear. Pronounced wear of these items and the grooves in the gearchange drum can lead to imprecise selection and, eventually, no selection. Renewal of the worn components will be required.
● Structural failure. Failure of any one component of the selector rod and change mechanism will result in improper or fouled gear selection.

25 Jumping out of gear

● Detent plunger assembly worn or damaged. Wear of the plunger and the cam with which it locates and breakage of the detent spring can cause imprecise gear selection resulting in jumping out of gear. Renew the damaged components.
● Gear pinion dogs worn or damaged. Rounding off the dog edges and the mating recesses in adjacent pinion can lead to jumping out of gear when under load. The gears should be inspected and renewed. Attempting to reprofile the dogs is not recommended.
● Selector forks, gearchange drum and pinion grooves worn. Extreme wear of these interconnected items can occur after high mileages especially when lubrication has been neglected. The worn components must be renewed.
● Gear pinions, bushes and shafts worn. Renew the worn components.
● Bent gearchange shaft. Often caused by dropping the machine on the gear lever.
● Gear pinion tooth broken. Chipped teeth are unlikely to cause jumping out of gear once the gear has been selected fully; a tooth which is completely broken off, however, may cause problems in this respect and in any event will cause transmission noise.

26 Overselection

● Claw arm spring weak or broken. Renew the spring.
● Detent plunger worn or broken. Renew the damaged items.
● Stopper arm spring worn or broken. Renew the spring.
● Gearchange arm stop pads worn. Repairs can be made by welding and reprofiling with a file.

Abnormal engine noise

27 Knocking or pinking

● See Section 16.

28 Piston slap or rattling from cylinder

● Cylinder bore/piston clearance excessive. Resulting from wear, or partial seizure. This condition can often be heard as a high, rapid tapping noise when the engine is under little or no load, particularly when power is just beginning to be applied. Reboring to the next correct oversize should be carried out and a new oversize piston fitted.
● Connecting rod bent. This can be caused by over-revving, trying to start a very badly flooded engine (resulting in a hydraulic lock in the cylinder) or by earlier mechanical failure. Attempts at straightening a bent connecting rod are not recommended. Careful inspection of the crankshaft should be made before renewing the damaged connecting rod.
● Gudgeon pin, piston boss bore or small-end bearing wear or seizure. Excess clearance or partial seizure between normal moving parts of these items can cause continuous or intermittent tapping noises. Rapid wear or seizure is caused by lubrication starvation.
● Piston rings worn, broken or sticking. Renew the rings after careful inspection of the piston and bore.

29 Other noises

● Big-end bearing wear. A pronounced knock from within the crankcase which worsens rapidly is indicative of big-end bearing failure as a result of extreme normal wear or lubrication failure. Remedial action in the form of a bottom end overhaul should be taken;

continuing to run the engine will lead to further damage including the possibility of connecting rod breakage.

● Main bearing failure. Extreme normal wear or failure of the main bearings is characteristically accompanied by a rumble from the crankcase and vibration felt through the frame and footrests. Renew the worn bearings and carry out a very careful examination of the crankshaft.

● Crankshaft excessively out of true. A bent crank may result from over-revving or damage from an upper cylinder component or gearbox failure. Damage can also result from dropping the machine on either crankshaft end. Straightening of the crankshaft is not be possible in normal circumstances; a replacement item should be fitted.

● Engine mounting loose. Tighten all the engine mounting nuts and bolts.

● Cylinder head gasket leaking. The noise most often associated with a leaking head gasket is a high pitched squeaking, although any other noise consistent with gas being forced out under pressure from a small orifice can also be emitted. Gasket leakage is often accompanied by oil seepage from around the mating joint or from the cylinder head holding down bolts and nuts. Leakage results from insufficient or uneven tightening of the cylinder head fasteners, or from random mechanical failure. Retightening to the correct torque figure will, at best, only provide a temporary cure. The gasket should be renewed at the earliest opportunity.

● Exhaust system leakage. Popping or crackling in the exhaust system, particularly when it occurs with the engine on the overrun, indicates a poor joint either at the cylinder port or at the exhaust pipe/silencer connection. Failure of the gasket or looseness of the clamp should be looked for.

Abnormal transmission noise

30 Clutch noise

● Clutch outer drum/friction plate tang clearance excessive.
● Clutch outer drum/spacer clearance excessive.
● Clutch outer drum/thrust washer clearance excessive.
● Primary drive gear teeth worn or damaged.
● Clutch shock absorber assembly worn or damaged.

31 Transmission noise

● Bearing or bushes worn or damaged. Renew the affected components.

● Gear pinions worn or chipped. Renew the gear pinions.

● Metal chips jammed in gear teeth.This can occur when pieces of metal from any failed component are picked up by a meshing pinion. The condition will lead to rapid bearing wear or early gear failure.

● Engine/transmission oil level too low. Top up immediately to prevent damage to gearbox and engine.

● Gearchange mechanism worn or damaged. Wear or failure of certain items in the selection and change components can induce mis-selection of gears (see Section 24) where incipient engagement of more than one gear set is promoted. Remedial action, by the overhaul of the gearbox, should be taken without delay.

● Chain snagging on cases or cycle parts. A badly worn chain or one that is excessively loose may snag or smack against adjacent components.

Exhaust smokes excessively

32 White/blue smoke (caused by oil burning)

● Cylinder cracked, worn or scored. These conditions may be caused by overheating, lack of lubrication, component failure or advanced normal wear. The cylinder barrel should be renewed and, if necessary, a new piston fitted.

● Petrol/oil ratio incorrect. Ensure that oil is mixed with the petrol in the correct ratio (machines converted to petroil lubrication only).

● Oil pump settings incorrect. Check and reset the oil pump as described in Routine Maintenance.

● Crankshaft main bearing oil seals worn. Wear in the main bearing

oil seals, often in conjunction with wear in the bearings themselves, can allow transmission oil to find its way into the crankcase and thence to the combustion chamber. This condition is often indicated by a mysterious drop in the transmission oil level with no sign of external leakage.

● Accumulated oil deposits in exhaust system. If the machine is used for short journeys only it is possible for the oil residue in the exhaust gases to condense in the relatively cool silencer. If the machine is then taken for a longer run in hot weather, the accumulated oil will burn off producing ominous smoke from the exhaust.

33 Black smoke (caused by over-rich mixture)

● Air filter element clogged. Clean or renew the element.

● Main jet loose or too large. Remove the float chamber to check for tightness of the jet. If the machine is used at high altitudes rejetting will be required to compensate for the lower atmospheric pressure.

● Cold start mechanism jammed on. Check that the mechanism works smoothly and correctly.

● Fuel level too high. The fuel level is controlled by the float height which can increase as a result of wear and damage. Remove the float bowl and check the float height. Check also that floats have not punctured; a punctured float will lose buoyancy and allow an increased fuel level.

● Float valve needle stuck open. Caused by dirt or a worn valve. Clean the float chamber or renew the needle and, if necessary, the valve seat.

Poor handling or roadholding

34 Directional instability

● Steering head bearing adjustment too tight. This will cause rolling or weaving at low speeds. Re-adjust the bearings.

● Steering head bearing worn or damaged. Correct adjustment of the bearing will prove impossible to achieve if wear or damage has occurred. Inconsistent handling will occur including rolling or weaving at low speed and poor directional control at indeterminate higher speeds. The steering head bearing should be dismantled for inspection and renewed if required. Lubrication should also be carried out.

● Bearing races pitted or dented. Impact damage caused, perhaps, by an accident or riding over a pot-hole can cause indentation of the bearing, usually in one position. This should be noted as notchiness when the handlebars are turned. Renew and lubricate the bearings.

● Steering stem bent. This will occur only if the machine is subjected to a high impact such as hitting a curb or a pot-hole. The bottom yoke/stem should be renewed; do not attempt to straighten the stem.

● Front or rear tyre pressures too low.

● Front or rear tyre worn. General instability, high speed wobbles and skipping over white lines indicates that tyre renewal may be required. Tyre induced problems, in some machine/tyre combinations, can occur even when the tyre in question is by no means fully worn.

● Swinging arm bearings worn. Difficulties in holding line, particularly when cornering or when changing power settings indicates wear in the swinging arm bearings. The swinging arm should be removed from the machine and the bearings renewed.

● Swinging arm flexing. The symptoms given in the preceding paragraph will also occur if the swinging arm fork flexes badly. This can be caused by structural weakness as a result of corrosion, fatigue or impact damage, or because the rear wheel spindle is slack.

● Wheel bearings worn. Renew the worn bearings.

● Loose wheel spokes. The spokes should be tightened evenly to maintain tension and trueness of the rim.

● Tyres unsuitable for machine. Not all available tyres will suit the characteristics of the frame and suspension, indeed, some tyres or tyre combinations may cause a transformation in the handling characteristics. If handling problems occur immediately after changing to a new tyre type or make, revert to the original tyres to see whether an improvement can be noted. In some instances a change to what are, in fact, suitable tyres may give rise to handling deficiences. In this case a thorough check should be made of all frame and suspension items which affect stability.

35 Steering bias to left or right

● Rear wheel out of alignment. Caused by uneven adjustment of chain tensioner adjusters allowing the wheel to be askew in the fork ends. A bent rear wheel spindle will also misalign the wheel in the swinging arm.
● Wheels out of alignment. This can be caused by impact damage to the frame, swinging arm, wheel spindles or front forks. Although occasionally a result of material failure or corrosion it is usually as a result of a crash.
● Front forks twisted in the steering yokes. A light impact, for instance with a pot-hole or low curb, can twist the fork legs in the steering yokes without causing structural damage to the fork legs or the yokes themselves. Re-alignment can be made by loosening the yoke pinch bolts, wheel spindle and mudguard bolts. Re-align the wheel with the handlebars and tighten the bolts working upwards from the wheel spindle. This action should be carried out only when there is no chance that structural damage has occurred.

36 Handlebar vibrates or oscillates

● Tyres worn or out of balance. Either condition, particularly in the front tyre, will promote shaking of the fork assembly and thus the handlebars. A sudden onset of shaking can result if a balance weight is displaced during use.
● Tyres badly positioned on the wheel rims. A moulded line on each wall of a tyre is provided to allow visual verification that the tyre is correctly positioned on the rim. A check can be made by rotating the tyre; any misalignment will be immediately obvious.
● Wheels rims warped or damaged. Inspect the wheels for runout as described in Chapter 5.
● Swinging arm bearings worn. Renew the bearings.
● Wheel bearings worn. Renew the bearings.
● Steering head bearings incorrectly adjusted. Vibration is more likely to result from bearings which are too loose rather than too tight. Re-adjust the bearings.
● Loosen fork component fasteners. Loose nuts and bolts holding the fork legs, wheel spindle, mudguards or steering stem can promote shaking at the handlebars. Fasteners on running gear such as the forks and suspension should be check tightened occasionally to prevent dangerous looseness of components occurring.
● Engine mounting bolts loose. Tighten all fasteners.

37 Poor front fork performance

● Damping fluid level incorrect. If the fluid level is too low poor suspension control will occur resulting in a general impairment of roadholding and early loss of tyre adhesion when cornering and braking. Too much oil is unlikely to change the fork characteristics unless severe overfilling occurs when the fork action will become stiffer and oil seal failure may occur.
● Damping oil viscosity incorrect. The damping action of the fork is directly related to the viscosity of the damping oil. The lighter the oil used, the less will be the damping action imparted. For general use, use the recommended viscosity of oil, changing to a slightly higher or heavier oil only when a change in damping characteristic is required. Overworked oil, or oil contaminated with water which has found its way past the seals, should be renewed to restore the correct damping performance and to prevent bottoming of the forks.
● Damping components worn or corroded. Advanced normal wear of the fork internals is unlikely to occur until a very high mileage has been covered. Continual use of the machine with damaged oil seals which allows the ingress of water, or neglect, will lead to rapid corrosion and wear. Dismantle the forks for inspection and overhaul.
● Weak fork springs. Progressive fatigue of the fork springs, resulting in a reduced spring free length, will occur after extensive use. This condition will promote excessive fork dive under braking, and in its advanced form will reduce the at-rest extended length of the forks and thus the fork geometry. Renewal of the springs as a pair is the only satisfactory course of action.
● Bent stanchions or corroded stanchions. Both conditions will prevent correct telescoping of the fork legs, and in an advanced state can cause sticking of the fork in one position. In a mild form corrosion

will cause stiction of the fork thereby increasing the time the suspension takes to react to an uneven road surface. Bent fork stanchions should be attended to immediately because they indicate that impact damage has occurred, and there is a danger that the forks will fail with disastrous consequences.

38 Front fork judder when braking (see also Section 46)

● Wear between the fork stanchions and the fork legs. Renewal of the affected components is required.
● Slack steering head bearings. Re-adjust the bearings.
● Warped brake drum. If irregular braking action occurs fork judder can be induced in what are normally serviceable forks. Renew the damaged brake components.

39 Poor rear suspension performance

● Rear suspension unit damper worn out or leaking. The damping performance of most rear suspension units falls off with age. This is a gradual process, and thus may not be immediately obvious. Indications of poor damping include hopping of the rear end when cornering or braking, and a general loss of positive stability.
● Weak rear springs. If the suspension unit springs fatigue they will promote excessive pitching of the machine and reduce the ground clearance when cornering. Although replacement springs are available separately from the rear suspension damper unit it is probable that if spring fatigue has occurred the damper units will also require renewal. See Sections 34 and 36.
● Swinging arm flexing or bearings worn.
● Bent suspension unit damper rod. This is likely to occur only if the machine is dropped or if seizure of the piston occurs. If either happens the suspension units should be renewed as a pair.

Abnormal frame and suspension noise

40 Front end noise

● Oil level low or too thin. This can cause a 'spurting' sound and is usually accompanied by irregular fork action.
● Spring weak or broken. Makes a clicking or scraping sound. Fork oil will have a lot of metal particles in it.
● Steering head bearings loose or damaged. Clicks when braking. Check, adjust or replace.
● Fork clamps loose. Make sure all fork clamp pinch bolts are tight.
● Fork stanchion bent. Good possibility if machine has been dropped. Repair or replace tube.

41 Rear suspension noise

● Fluid level too low. Leakage of a suspension unit, usually evident by oil on the outer surfaces, can cause a spurting noise. The suspension units should be renewed as a pair.
● Defective rear suspension unit with internal damage. Renew the suspension units as a pair.

Brake problems

42 Brakes are spongy or ineffective

● Brake cable deterioration. Damage to the outer cable by stretching or being trapped will give a spongy feel to the brake lever. The cable should be renewed. A cable which has become corroded due to old age or neglect of lubrication will partially seize making operation very heavy. Lubrication at this stage may overcome the problem but the fitting of a new cable is recommended.
● Worn brake linings. Determine lining wear using the external brake wear indicator on the brake backplate, or by removing the wheel and withdrawing the brake backplate. Renew the shoe/lining units as a pair if the linings are worn below the recommended limit.
● Worn brake camshaft. Wear between the camshaft and the

bearing surface will reduce brake feel and reduce operating efficiency. Renewal of one or both items will be required to rectify the fault.
● Worn brake cam and shoe ends. Renew the worn components.
● Linings contaminated with dust or grease. Any accumulations of dust should be cleaned from the brake assembly and drum using a petrol dampened cloth. Do not blow or brush off the dust because it is asbestos based and thus harmful if inhaled. Light contamination from grease can be removed from the surface of the brake linings using a solvent; attempts at removing heavier contamination are less likely to be successful because some of the lubricant will have been absorbed by the lining material which will severely reduce the braking performance.

43 Brake drag

● Incorrect adjustment. Re-adjust the brake operating mechanism.
● Drum warped or oval. This can result from overheating or impact or uneven tension of the wheel spokes. The condition is difficult to correct, although if slight ovality only occurs, skimming the surface of the brake drum can provide a cure. This is work for a specialist engineer. Renewal of the complete wheel hub is normally the only satisfactory solution.
● Weak brake shoe return springs. This will prevent the brake lining/shoe units from pulling away from the drum surface once the brake is released. The springs should be renewed.
● Brake camshaft, lever pivot or cable poorly lubricated. Failure to attend to regular lubrication of these areas will increase operating resistance which, when compounded, may cause tardy operation and poor release movement.

44 Brake lever or pedal pulsates in operation

● Drums warped or oval. This can result from overheating or impact or uneven spoke tension. This condition is difficult to correct, although if slight ovality only occurs skimming the surface of the drum can provide a cure. This is work for a specialist engineer. Renewal of the hub is normally the only satisfactory solution.

45 Drum brake noise

● Drum warped or oval. This can cause intermittent rubbing of the brake linings against the drum. See the preceding Section.
● Brake linings glazed. This condition, usually accompanied by heavy lining dust contamination, often induces brake squeal. The surface of the linings may be roughened using glass-paper or a fine file.

46 Brake induced fork judder

● Worn front fork stanchions and legs, or worn or badly adjusted steering head bearings. These conditions, combined with uneven or pulsating braking as described in Section 44 will induce more or less judder when the brakes are applied, dependent on the degree of wear and poor brake operation. Attention should be given to both areas of malfunction. See the relevant Section.

Electrical problems

47 Battery dead or weak

● Battery faulty. Battery life should not be expected to exceed 3 to 4 years. Gradual sulphation of the plates and sediment deposits will reduce the battery performance. Plate and insulator damage can often

occur as a result of vibration. Complete power failure, or intermittent failure, may be due to a broken battery terminal. Lack of electrolyte will prevent the battery maintaining charge.
● Battery leads making poor contact. Remove the battery leads and clean them and the terminals, removing all traces of corrosion and tarnish. Reconnect the leads and apply a coating of petroleum jelly to the terminals.
● Load excessive. If additional items such as spot lamps, are fitted, which increase the total electrical load above the maximum generator output, the battery will fail to maintain full charge. Reduce the electrical load to suit the electrical capacity.
● Rectifier failure.
● Alternator generating coils open-circuit or shorted.
● Charging circuit shorting or open circuit. This may be caused by frayed or broken wiring, dirty connectors or a faulty ignition switch. The system should be tested in a logical manner. See Section 50.

48 Battery overcharged

● Rectifier faulty. Overcharging is indicated if the battery becomes hot or it is noticed that the electrolyte level falls repeatedly between checks. In extreme cases the battery will boil causing corrosive gases and electrolyte to be emitted through the vent pipes.
● Battery wrongly matched to the electrical circuit. Ensure that the specified battery is fitted to the machine.

49 Total electrical failure

● Fuse blown. Check the main fuse. If a fault has occurred, it must be rectified before a new fuse is fitted.
● Battery faulty. See Section 47.
● Earth failure. Check that the frame main earth strap from the battery is securely affixed to the frame and is making a good contact.
● Ignition switch or power circuit failure. Check for current flow through the battery positive lead (red) to the ignition switch. Check the ignition switch for continuity.

50 Circuit failure

● Cable failure. Refer to the machine's wiring diagram and check the circuit for continuity. Open circuits are a result of loose or corroded connections, either at terminals or in-line connectors, or because of broken wires. Occasionally, the core of a wire will break without there being any apparent damage to the outer plastic cover.
● Switch failure. All switches may be checked for continuity in each switch position, after referring to the switch position boxes incorporated in the wiring diagram for the machine. Switch failure may be a result of mechanical breakage, corrosion or water.
● Fuse blown. Refer to the wiring diagram to check whether or not a circuit fuse is fitted. Replace the fuse, if blown, only after the fault has been identified and rectified.

51 Bulbs blowing repeatedly

● Vibration failure. This is often an inherent fault related to the natural vibration characteristics of the engine and frame and is, thus, difficult to resolve. Modifications of the lamp mounting, to change the damping characteristics, may help.
● Intermittent earth. Repeated failure of one bulb, particularly where the bulb is fed directly from the generator, indicates that a poor earth exists somewhere in the circuit. Check that a good contact is available at each earthing point in the circuit.
● Reduced voltage. Do not overload the system with additional electrical equipment in excess of the system's power capacity and ensure that all circuit connections are maintained clean and tight.

YAMAHA TY 50, 80, 125 & 175

Check list

Note: *for off-road use, see maintenance schedules listed at the beginning of the Routine maintenance section in manual.*

Pre-riding check (daily)

1 Check the engine oil level
2 Check the petrol level
3 Check the operation of the brakes
4 Inspect the tyres for damage and check the pressures
5 Ensure that the final drive chain is lubricated and correctly adjusted
6 Check the controls and steering for correct operation
7 Ensure that the electrical system and speedometer function correctly

Two weekly, or every 300 miles (500 km)

1 Check, adjust and lubricate the final drive chain

Six weekly, or every 1000 miles (1500 km)

1 Clean the air filter element
2 Check the transmission oil level
3 Overhaul and adjust the brakes

Three monthly, or every 2000 miles (3000 km)

1 Decarbonise the engine and exhaust system
2 Check the contact breaker points and ignition timing
3 Check the spark plug
4 Clean the fuel tap
5 Oil the control cables and grease the twistgrip, stand pivot and controls
6 Check the carburettor, throttle cable and fuel pipe
7 Check the oil pump settings
8 Adjust the clutch
9 Change the transmission oil
10 Check the battery condition and electrolyte level
11 Check the steering head bearings and the suspension
12 Check the wheels and tyres
13 Check all fittings, fasteners, lights and signals

Six monthly, or every 4000 miles (6000 km)

1 Overhaul the carburettor
2 Grease the rear suspension pivot
3 Change the front fork oil

Annually, or every 8000 miles (12 000 km)

1 Grease the steering head bearings
2 Overhaul the brakes
3 Grease the wheel bearings and speedometer drive

Adjustment data

Spark plug gap
TY50, 80 and 175   0.5 – 0.6 mm (0.020 – 0.024 in)
TY125   0.7 – 0.8 mm (0.028 – 0.032 in)

Spark plug type
TY50   NGK B7HS
TY80   NGK B6HS
TY125 and 175   NGK B7ES

Contact breaker gap   0.35 mm (0.014 in)

Ignition timing – BTDC 1.80 ± 0.15 mm (0.0709 ± 0.0059 in)

Idle speed
TY50 and 80 1300 ± 50 rpm
TY125 and 175 1100 ± 50 rpm

Tyre pressures	Front	Rear
TY50 P and early TY50 M	21 psi	29 psi
Late TY50 M	20 psi	29 psi
TY80	26 psi	29 psi
TY125 and 175	13 psi	16 psi

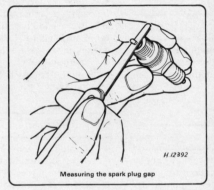

Measuring the spark plug gap

H.12392

Recommended lubricants

Component	Quantity	Type/viscosity
❶ Engine:		Good quality, self-mixing two-stroke oil
TY50	1.20 lit (2.11 pt)	
TY80	0.22 lit (0.39 pt)	
TY125 and 175	0.30 lit (0.53 pt)	
❷ Gearbox:		SAE 10W/30 SE engine oil
TY50 and 80	500 – 550 cc (0.88 – 0.97 pt)	
TY125 and 175	650 cc (1.14 pt)	
❸ Front forks – see manual for recommended quantities and oils		
❹ Final drive chain	As required	Chain grease
❺ Air filter	As required	SAE 30W engine oil
❻ Wheel bearings	As required	High melting point grease
❼ Brake camshafts	As required	High melting point grease
❽ Steering head bearings	As required	Multi-purpose grease
❾ Swinging arm pivot bearings	As required	Multi-purpose grease
❿ Cables	As required	Light machine oil
⓫ Junction box	As required	Dry graphite-based lubricant

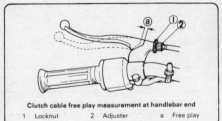

Clutch cable free play measurement at handlebar end

1 Locknut   2 Adjuster   a Free play

ROUTINE MAINTENANCE GUIDE

Routine maintenance

Periodic routine maintenance is a continuous process which should commence immediately the machine is used. The object is to maintain all adjustments and to diagnose and rectify minor defects before they develop into more extensive, and often more expensive, problems.

It follows that if the machine is maintained properly, it will both run and perform with optimum efficiency, and be less prone to unexpected breakdowns. Regular inspection of the machine will show up any parts which are wearing, and with a little experience, it is possible to obtain the maximum life from any one component, renewing it when it becomes so worn that it is liable to fail.

Note that some special tools are required for routine maintenance, also that a good selection of general workshop tools is essential. Included in the tools must be a range of metric ring or combination spanners, a selection of crosshead screwdrivers, and two pairs of circlip pliers, one external opening and the other internal opening. Additionally, owing to the extreme tightness of most casing screws on Japanese machines, an impact screwdriver, together with a choice of large or small cross-head screw bits, is absolutely indispensable. This is particularly so if the engine has not been dismantled since leaving the factory. One further item of equipment that can be regarded as essential is a stand of some sort that can hold the machine securely in an upright position with enough height to permit the wheels to be removed. This can be anything from an old milk crate to a purpose-built paddock-type metal stand.

Maintenance schedules

All tasks are grouped under various mileage headings, all of which are also given calendar-based intervals; if the machine only covers a low mileage maintenance should be carried out according to the calendar headings instead. All intervals are intended as a guide only; as a machine gets older it develops individual faults which require more frequent attention and if used under particularly arduous conditions it is advisable to reduce the period between each check.

This is particularly true in the case of machines used off-road; first, the usual lack of a speedometer means that the mileage headings can no longer be followed; secondly, the severity of the conditions in which most off-road machines are used means that the largely road-use orientated calendar headings are far too infrequent.

While it is impossible to give a maintenance schedule applicable to all off-road machines, individual usage being subject to far too many variables, it is possible to give a few general recommendations which can be followed and modified by the individual owner as and when he or she gains the experience and the knowledge of his or her own machine's requirements.

General recommendations

1   All the tasks listed in the Routine Maintenance section of this manual should be carried out in the form of a major overhaul at least once each season, more often if the machine is used regularly in competition.
2   Cut the recommended mileage/time intervals by at least half; if no speedometer is fitted, estimate mileage covered on the basis of a 10 – 15 mph average speed.

Before each competition:

Carry out the full pre-riding check given, paying particular attention to lubricating all controls and the drive chain. Clean the air filter, grease the swinging-arm pivot (where applicable) and test-ride the machine to ensure that all is well.

After each competition:

1   Clean the machine thoroughly, removing all mud and dirt while it is still wet, then take it for a short run to dry it properly.
2   Remove the chain for thorough cleaning and greasing in chain grease.
3   Remove the flywheel generator (and oil pump, if applicable) cover to remove all traces of water or of moisture caused by condensation. Check the points gap and ignition timing, apply one or two drops of oil to the cam lubricating wick (and check the oil pump settings, if applicable), then refit the cover(s) using new gaskets, if required.
4   Remove the spark plug, check its gap or renew it as necessary, check the HT coil, lead and suppressor cap.
5   Remove the carburettor float bowl drain plug and the fuel tap filter bowl, collecting the petrol that flows out in a clean container; if any sign of water or dirt is seen, clean out the carburettor and tank.
6   Clean the air filter.
7   Check that all control cables are correctly adjusted, properly routed and protected against the entry of dirt or water, and that all are lubricated.
8   Remove each wheel in turn to dry out and clean the brakes, check the bearings (greasing or renewing them as necessary), the spokes and wheel rims and the tyres. If the tyre leading edges are worn, reverse the tyres on the rims to present a fresh surface for the next competition.
9   Check the transmission oil level.
10   Repair any damage to the machine. Note that if all cycle parts are given a good coating of wax polish, the machine will be much easier to clean in the future, apart from preserving its appearance.
11   Test ride the machine to ensure it is completely ready for the next event.

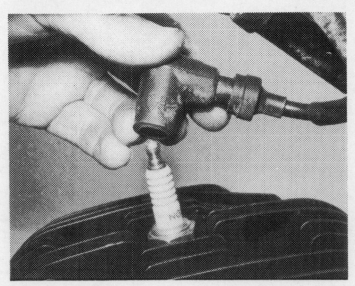

Check carefully spark plug and HT lead – renew if necessary

Note that some components will require careful work by an expert if repairs are necessary

After every two competitions:
1   Strip the engine top end, check for wear, decoke, renew the piston rings if worn and reassemble.
2   Decoke the exhaust system.
3   Overhaul the carburettor.
4   Check ignition timing.
5   Change the transmission oil.
6   Change the fork oil and remove any dirt from under the dust covers, above the seals.
7   Check the steering head bearings and swinging arm pivot bearing – dismantle and pack with grease if necessary.

Cleaning the machine
Regular cleaning can be considered as important as mechanical maintenance. This will ensure that all the cycle parts are inspected regularly and are kept free from accumulations of road dirt and grime.

Cleaning is especially important during the winter months or during the trials season, despite its appearance of being a thankless task which very soon seems pointless. On the contrary, it is at this time that the paintwork, chromium plating, and the alloy casings suffer the ravages of abrasive grit, rain and road salt. A couple of hours spent weekly on cleaning the machine will maintain its appearance and value, and highlight small points, like chipped paint, before they become a serious problem.

Use a sponge and copious amounts of warm soapy water to wash surface dirt from these components. Remove oil and grease with a solvent such as 'Gunk' or 'Jizer', working it in with a stiff brush when the component is still dry and rinsing it off with fresh water. Keep water out of the carburettor, air filter and electrics.

Apply wax polish to the painted components and to those which are chromed. Keep the chain and control cables well lubricated to prevent the ingress of water and wipe the machine down if used in the wet.

Do not use strong detergents, scouring powders or any abrasive when cleaning plastic components; anything but a mild solution of soapy water may well bleach or score the surface. On completion of cleaning, wipe the component dry with a chamois leather. If the surface finish has faded, use a fine aerosol polish to restore its shine.

Note: while it is realised that cleaning a machine that is plastered in mud and dirt after a trial is quickest and most effective if carried out using a pressure washer, steam cleaner or even a very powerful hose, the very real disadvantages of such usage should be pointed out. Quite apart from the rapid deterioration of the finish of plastic components and of all painted or lacquered metal components caused by the scouring action of caked-on dirt being blasted off, the operating pressure of such machines is high enough to force a mixture of dirt and water past oil seals etc and into the bearings, brakes, forks and suspension units, causing their premature failure unless great care is taken to dismantle, clean and lubricate all cycle parts after cleaning. If cleaning must be carried out in this way, be very careful both when cleaning and afterwards.

Pre-riding check (daily)
Before taking the machine out on the road there are certain checks which should be completed to ensure that it is in a safe and legal condition to be used.

1   Engine oil level check
Although it is safe to use the machine as long as oil is visible in the tank sight glass, it is recommended that the level is maintained to within approximately an inch of the filler hole to allow a good reserve. Detach the left-hand sidepanel or on TY80 models, the seat, to expose the filler cap and top up using only a good quality two-stroke engine oil. It is useful to keep a spare container of oil with the machine so that a supply is always available.

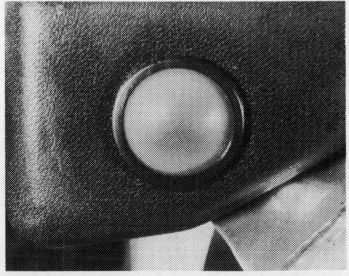

Oil level must not be allowed to fall below bottom edge of tank sight glass

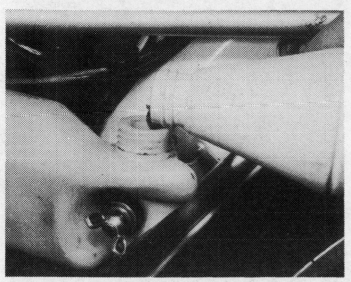

Use only good quality two-stroke engine oil when topping up

Lock cap on to valve stem as shown to provide early warning of tyre creep during trials use

2 Petrol level

Checking the petrol level may seem obvious, but it is all too easy to forget. Ensure that you have enough petrol to complete your journey, or at least to get you to the nearest petrol station.

3 Brakes

Check that the front and rear brakes work effectively and without binding. Ensure that the rod linkages and the cables, as applicable, are lubricated and properly adjusted.

4 Tyres

Check the tyre pressure with a gauge that is known to be accurate. It is worthwhile purchasing a pocket gauge for this purpose because the gauges on garage forecourt airlines are notoriously inaccurate. The pressures should be checked with the tyres cold. Even a few miles travelled will warm up the tyres to a point where pressures increase and an inaccurate reading will result.

At the same time as the tyre pressures are checked, examine the tyres themselves. Check them for damage, especially splitting of the sidewalls. Remove any small stones or other road debris caught between the treads. This is particularly important on the rear tyre, where rapid deflation due to penetration of the inner tube will almost certainly cause total loss of control. When checking the tyres for damage, they should be examined for tread depth in view of both the legal and safety aspects. It is vital to keep the tread depth within the UK legal limits of 1 mm of depth over three-quarters of the tread breadth around the entire circumference with no sign of bald patches. Many riders, however, consider nearer 2 mm to be the limit for secure roadholding, traction, and braking, especially in adverse weather conditions. Obviously this applies only to machines used on the road; trials riders use a completely different set of standards to ensure that they have maximum grip in competitions. Expert riders renew their rear tyres when the leading edges of all knobs are rounded off (ie, after two events). Front tyres generally last longer, being renewed with every third rear tyre; a lot will depend on how much money the rider has to spend and what standard he or she has reached but no trials tyre should be used when its knobs have worn to 4 mm or less. Tyre pressures should be chosen and set at the beginning of each event when the actual conditions can be seen. Do not forget to check that all security bolts are tight. Valve caps should be locked on to the stem with the locking ring as shown in the accompanying photograph so that if the tyre does begin to creep on the rim and pull the tube with it this can be detected by the angle of the valve stem and rectified before the stem is torn out of the tube.

5 Final drive chain

Check that the final drive chain is correctly adjusted and well lubricated. Remember that if the machine is used in adverse conditions the chain will require frequent, even daily, lubrication. Refer to the 2 weekly/300 mile service interval.

6 Controls and steering

Check throttle, clutch, gear lever and footrests to ensure that they are securely fastened and working properly. If a bolt is going to work loose, or a cable snap, it is better that it does so with the machine at a standstill than when riding. Check also that the steering and suspension are working correctly.

7 Lights and speedometer

Check that all lights, turn signals, horn and speedometer are working correctly to make sure that the machine complies with all legal requirements in this respect.

Two weekly, or every 300 miles (500 km)

This is where the correct routine maintenance procedure begins. The daily checks serve to ensure that the machine is safe and legal to use, but contribute little to maintenance other than to give the owner an accurate picture of what item needs attention. However, if done conscientiously they will give early warning of any faults which are about to appear. When performing the following maintenance tasks, therefore, carry out the daily checks first.

1 Check, adjust and lubricate the final drive chain

The chain consists of a multitude of small bearing surfaces which will wear rapidly, and expensively, if the chain is not regularly lubricated and adjusted. A simple check for wear is as follows. With the chain fully lubricated and correctly adjusted as described below, attempt to pull the chain backwards off the rear sprocket. If the chain can be pulled clear of the sprocket teeth it must be considered worn out and renewed; chains should be renewed always in conjunction with the sprockets since the running together of new and part-worn components will greatly increase the rate of wear of both, necessitating renewal much sooner than would otherwise be the case.

A more accurate measurement of chain wear involves the removal of the chain from the machine and its thorough cleaning. Disconnect the chain at its split connecting link and pull the entire length of the chain clear of the sprockets. Note that refitting the chain is greatly simplified if a worn out length is temporarily connected to it. As the original chain is pulled off the sprockets, the worn-out chain will follow

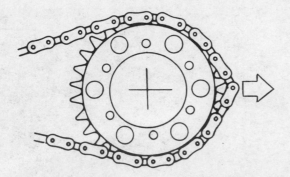

Checking for an excessively worn final drive chain

it and remain in place while the task of cleaning and examination is carried out. On reassembly, the process is repeated, pulling the worn-out chain over the sprockets so that the new chain, or the freshly cleaned and lubricated chain, is pulled easily into place.

To clean the chain, immerse it in a bath containing a mixture of petrol and paraffin and use a stiff-bristled brush to scrub away all the traces of road dirt and old lubricant. Take the necessary fire precautions when using this flammable solvent. Swill the chain around to ensure that the solvent penetrates fully into the bushes and rollers and can remove any lubricant which may still be present. When the chain is completely clean, remove it from the bath and hang it up to dry.

To access accurately the amount of wear present in the chain, it must be cleaned and dried as described above, then laid out on a flat surface. Compress the chain fully and measure its length from end to end. Anchor one end of the chain and pull on the other end, drawing the chain out to its fullest extent. Measure the stretched length. If the stretched measurement exceeds the compressed measurement by move than $\frac{1}{4}$ in per foot, the chain must be considered worn out and be renewed.

Chain lubrication is best carried out by immersing the chain in a molten lubricant such as Chainguard or Linklyfe. Lubrication carried out in this manner must be preceded by removing the chain from the machine, cleaning it, and drying it as described above. Follow the manufacturer's instructions carefully when using Chainguard or Linklyfe, and take great care to swill the chain gently in the molten lubricant to ensure that all bearing surfaces are fully greased.

Refitting a new, or freshly-lubricated, chain is a potentially messy affair which is greatly simplified by connecting it to the worn-out length of chain used during its removal to pull it around the sprockets. Refit the connecting link, ensuring that the spring clip is fitted with its closed end facing the normal direction of travel of the chain.

For the purpose of daily or weekly lubrication, one of the many proprietary aerosol-applied chain lubricants can be applied, while the chain is in place on the machine. It should be applied at least once a week, and daily if the machine is used in wet weather conditions. If the roller surfaces look dry, then they need lubrication. Engine oil can be used for this task, but remember that it is flung off the chain far more easily than grease, thus making the rear end of the machine unnecessarily dirty, and requires more frequent application if it is to perform its task adequately. Also remember that surplus oil will eventually find its way on to the tyre, with quite disastrous consequences.

It is necessary to check the chain tension at regular intervals to compensate for wear. Since this wear does not take place evenly along the length of the chain, tight spots will appear which must be compensated for when adjustment is made. Chain tension is checked with the transmission in neutral and with the rider seated normally on the machine, which must be standing on its wheels. Find the tightest spot in the chain by pushing the machine along and feeling the amount of free play present on the bottom run of the chain, midway between the sprockets, testing along the entire length of the chain. Press the chain tensioner, where fitted, away from the chain to ensure that free play is measured accurately. When the tightest spot has been found, measure the total amount of up and down movement available; this should be approximately 30 mm (1.2 in) on TY175 models, 19 – 25 mm ($\frac{3}{4}$ – 1 in) on all others.

If adjustment is necessary, remove the split pin, slacken the rear wheel spindle nut just enough to permit the spindle to be moved, then draw the spindle back by rotating the snail cams. Use the marks stamped in each cam to ensure that the spindle is moved back by the same amount on each side, thus preserving accurate wheel alignment. TY50 models are fitted with adjuster drawbolts; turn each nut by exactly the same amount to preserve wheel alignment. As a guide to this ensure that the stamped reference mark on each adjuster is aligned with the same index line stamped in the swinging arm fork ends. A final check of accurate wheel alignment can be made by laying a plank of wood or drawing a length of string parallel to the machine so that it touches both walls of the rear tyre. Wheel alignment is correct when the plank or string is equidistant from both walls of the front tyre when tested on both sides of the machine, as shown in the accompanying illustration.

When the chain is correctly tensioned, apply the rear brake to centralise the shoes on the drum and tighten the spindle retaining nut securely. Use the recommended torque settings where possible and do not forget to fit a new split pin to prevent the nut from slackening.

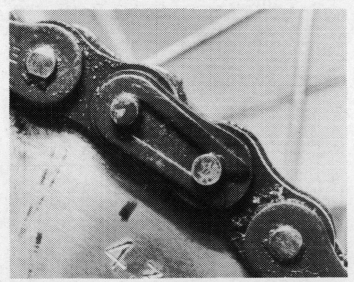

Spring clip must be refitted with closed end facing normal direction of chain travel, as shown

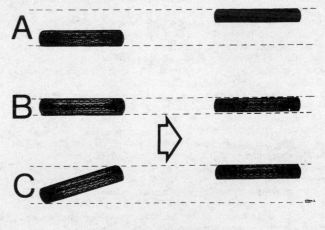

Checking the wheel alignment

A and C – Incorrect
B – Correct

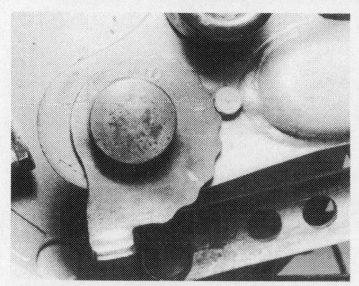

Ensure wheel alignment is correct by aligning the same cutout in each cam with the stop – stamped marks assist identification of cutouts

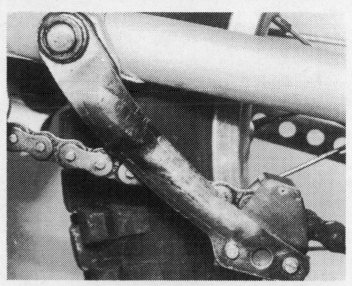

Check tensioner works correctly and renew block if worn excessively

Note that if the chain tension has been altered significantly, the rear brake and stop lamp rear switch adjustment will also require resetting; these should be checked before taking the machine out on the road.

Note that replacement chains are available in standard metric sizes from Renold Ltd, the British chain manufacturer. When purchasing a new chain always quote the size, the number of links required and the machine to which the chain is to be fitted. **Standard** chain sizes are given in the Specifications Section of Chapter 1, but note that the gearing may well have been altered on trials machines. It is a popular and worthwhile modification on TY175 models to fit a larger 520 ($\frac{5}{8}$ x $\frac{1}{4}$ in) chain with suitable sprockets to gain much longer sprocket and chain life at the expense of a slight weight increase.

Finally, on machines so equipped, check that the chain tensioner rubbing block is in good condition, renewing it if worn excessively, and that the tensioner arm moves freely against spring pressure. Apply a few drops of oil to the tensioner pivot and renew the spring if it appears to have weakened or if it is damaged.

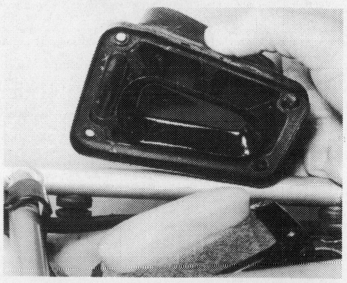

Remove the cover retaining screw(s) and lift away the air filter cover

Six weekly, or every 1000 miles (1500 km)

Complete the operations under the previous mileage/time heading, then carry out the following:

1 Clean the air filter element

Raise or remove the seat, or remove the left-hand sidepanel (as applicable) then remove the screw(s) retaining the filter cover, lift out the filter element and separate it from its supporting frame.

Renew the element if hardened or badly clogged. If serviceable, immerse the element in white spirit or a similar high-flash point solvent gently squeezing it to remove all oil and dirt. Remove excess solvent by pressing the element between the palms of the hands; wringing it out will cause damage. Allow a short time for any remaining solvent to evaporate.

Reimpregnate the element with clean SAE 30W oil and gently squeeze out any excess. Coat the element sealing edges with light grease and refit it to its frame.

When fitting, position the element and its cover correctly. Air which bypasses the element will carry dirt into the carburettor and crankcase and will weaken the fuel/air mixture; apply grease to the cover sealing lips for extra protection.

If riding in a particularly dusty or moist atmosphere, increase the frequency of cleaning the element. Never run the engine without the element fitted; the carburettor is jetted to compensate for its being fitted and the resulting weak mixture will cause overheating of the engine.

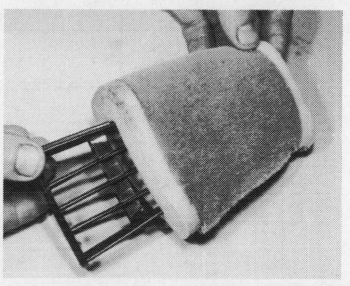

Detach foam element from its frame for cleaning and re-oiling

2   Check the transmission oil level

Start the engine and allow it to idle for a few minutes to warm the oil and to distribute it fully so that the true level is recorded. Stop the engine and remove the filler plug. Wipe the dipstick clean and insert it into the filler plug orifice. When the machine is standing upright on its wheels on level ground, the oil level should be between the 'Maximum' and 'Minimum' marks on the dipstick which should be positioned squarely on the mouth of the orifice; where the dipstick is integral with the filler plug it should be rested on the casing and **not** screwed in.

Add (or remove) oil if necessary; use only a good quality SAE 10W/30SE engine oil when topping up, and ensure that the filler plug is securely fastened when the level is correct.

3   Overhaul and adjust the brakes

The cable-operated front brake requires regular adjustment to compensate for shoe wear and for variations in the cable itself. To check that the adjustment is correct, apply the front brake firmly and measure the distance between the handlebar lever butt end and the lever clamp. The distance should be 5 – 8 mm (0.2 – 0.3 in) when the handlebar lever is firmly applied. If adjustment is necessary, it should be made at the adjuster set on or just above the brake backplate. Once the front brake has been correctly adjusted, spin the front wheel and apply the front brake hard to settle the cable and brake components. Check that the adjustment has not been altered, re-setting it if necessary, and ensure that the adjuster locknuts are securely tightened and that all the rubber cable protecting sleeves are correctly replaced. Check that the front wheel is free to rotate easily. The adjuster at the handlebar lever clamp should be reserved for quick roadside adjustment; be sure that it is screwed fully in before using the lower adjuster.

The rear brake is adjusted by means of a single nut at the rear end of the brake operating rod. Turn the nut clockwise to reduce free play, if necessary, to measure 20 – 30 mm (0.8 – 1.2 in) on TY125 models, at least 25 mm (1.0 in) on all others, at the brake pedal tip. Check that the rear wheel rotates freely and that the stop lamp is functioning properly. Remember that the stop lamp switch height must be adjusted every time the rear brake adjustment is altered. To adjust the switch height, turn its plastic sleeve nut as required until the stop lamp bulb lights when the brake pedal free play has been taken up and the rear brake shoes are just beginning to engage the brake drum.

Complete brake maintenance by oiling all lever pivot points, all exposed lengths of cable, cable nipples and the rear brake linkage with a few drops of oil from a can. Remember not to allow excessive oil onto the operating linkage, in case any surplus should find its way into the brake drum or onto the tyre.

Regular checks must be made to ensure that the friction material of the brake shoes is not worn down to a dangerous level, and to ensure that worn items are renewed in good time to maintain peak brake efficiency. This is easily checked with no need for dismantling. If, with the brake correctly adjusted and firmly applied, the angle formed between the operating lever on the brake backplate and the brake rod or cable is more than 90 degrees, the friction material is seriously worn and the shoes should be renewed. On some models, this can be checked by removing the rubber plug set in the brake backplate. If, on dismantling, the shoes are found to be within wear limits it is possible to remove the operating lever from the camshaft and to rotate it through one or two splines until the angle is less than 90 degrees, thus restoring the maximum efficiency of the operating mechanism.

Overhauling the brakes must be preceded by the removal of the wheel concerned, as described in the relevant Sections of Chapter 5. The brake components can then be dismantled, cleaned, checked for wear, and reassembled following the instructions given in the same Chapter. It is important that moving parts such as the brake camshaft are lubricated with a smear of high melting-point grease on reassembly. **Note**: the brakes must be overhauled at this interval regardless of the amount of friction material remaining.

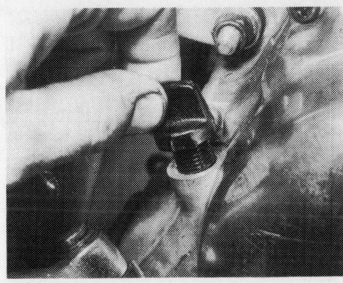

Dipstick should be positioned as shown to check transmission oil level – run engine as described before checking

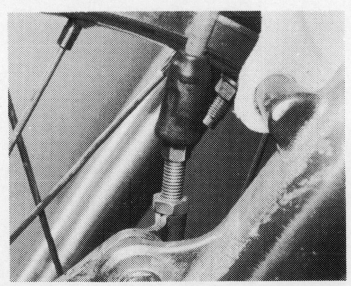

Front brake is adjusted at adjuster on fork lower leg – TY125 and 175

Rear brake is adjusted at rear end of operating rod

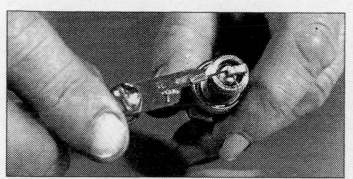

Electrode gap check - use a wire type gauge for best results

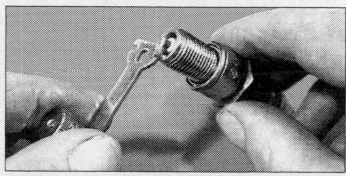

Electrode gap adjustment - bend the side electrode using the correct tool

Normal condition - A brown, tan or grey firing end indicates that the engine is in good condition and that the plug type is correct

Ash deposits - Light brown deposits encrusted on the electrodes and insulator, leading to misfire and hesitation. Caused by excessive amounts of oil in the combustion chamber or poor quality fuel/oil

Carbon fouling - Dry, black sooty deposits leading to misfire and weak spark. Caused by an over-rich fuel/air mixture, faulty choke operation or blocked air filter

Oil fouling - Wet oily deposits leading to misfire and weak spark. Caused by oil leakage past piston rings or valve guides (4-stroke engine), or excess lubricant (2-stroke engine)

Overheating - A blistered white insulator and glazed electrodes. Caused by ignition system fault, incorrect fuel, or cooling system fault

Worn plug - Worn electrodes will cause poor starting in damp or cold weather and will also waste fuel

Three monthly, or every 2000 miles (3000 km)

Complete the operations listed under the previous mileage/time headings, then carry out the following:

1 Decarbonise the engine and exhaust system

Cylinder head and barrel

Refer to Chapter 1 and remove the cylinder head and barrel. It is necessary to remove all carbon from the head, barrel and piston crown whilst avoiding removal of the metal surfaces on which it is deposited. Take care when dealing with the soft alloy head and piston. Never use a steel scraper or screwdriver. A hardwood, brass or aluminium scraper is ideal as these are harder than the carbon but not harder than the underlying metal. With the bulk of carbon removed, use a brass wire brush. Finish the head and piston with metal polish; a polished surface will slow the subsequent build-up of carbon. Clean out the barrel ports to prevent the restriction of gas flow. Remove all debris by washing each component in paraffin whilst observing the necessary fire precautions. Renew the piston rings, if necessary, on reassembly.

Exhaust system

Remove, where possible, the exhaust silencer baffle by removing its retaining screw with washer(s), gripping its end with pliers or a mole wrench and pulling it from position. If the baffle has seized in position, pass a tommy bar through the mole wrench and strike it rearwards with a hammer.

Remove and discard any wrapping around the baffle; it need not be replaced. If the build-up of carbon and oil on the baffle is not too great, wash it clean with petrol whilst taking the necessary fire precautions. Heavy deposits on the baffle may well indicate similar contamination within the silencer and pipe in which case the system should be removed. Clean the baffle by running a blowlamp along its length to burn off the deposits, waiting for it to cool and tapping it sharply with a length of hardwood to dislodge any remaining deposits. Finish with a wire brush and check the baffle holes are clear.

Suspend the two halves of the system each from its rearmost end. Block the lower end of each half with a cork or wooden bung ensuring enough protrudes for removal purposes. Mix up a caustic soda solution (3 lb to over a gallon of fresh water), adding the soda to the water gradually, whilst stirring. Do not pour water into a container of soda, this will cause a violent reaction to take place. Wear proper eye and skin protection; caustic soda is very dangerous. Eyes and skin contaminated by soda must be immediately flushed with fresh water and examined by a doctor. The solution will react violently with aluminium alloy, causing severe damage to any components.

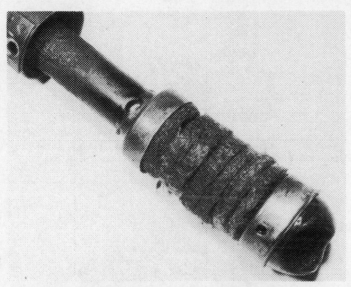

Wrapping should be removed from exhaust baffles ...

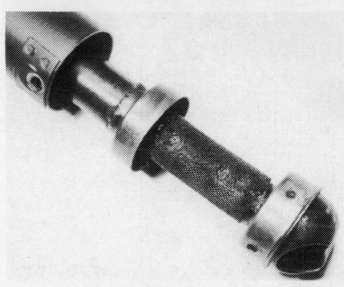

... so that they can be properly cleaned

Fill each half of the system with solution, leaving the upper end open. Leave the solution overnight to allow its dissolving action to take place. Ventilate the area to prevent the build-up of noxious fumes. On completion, carefully pour out the solution and flush the system through with clean, fresh water.

Refit the system. Where applicable. renew the O-ring sealing the joint between the two parts of the system. Lightly smear the baffle mating surface and retaining screw threads to prevent seizure. Renew the screw spring washer if flattened.

Do not modify the baffle or run the machine with it removed. This will result in less performance and affect the carburation.

2 Check the contact breaker points and ignition timing

Remove the left-hand crankcase cover. The contact breaker assembly can be viewed through one of the generator rotor aperture slots. Remove the spark plug and turn the rotor until the points begin to open. Use a small screwdriver to push the moving point open against its spring. Examine the point contact faces. If they are burnt or pitted, remove the points for cleaning or renewal. Light surface deposits can be removed with crocus paper or a piece of stiff card.

To remove the points, lock the crankshaft by selecting top gear and applying the rear brake hard or by applying a holding tool to the rotor itself, as shown in the accompanying photograph, then remove the rotor retaining nut (and washers) and withdraw the rotor using a flywheel puller as described in Chapter 1. Remove the contact breaker

Always renew exhaust sealing gaskets and O-rings to prevent leaks upsetting smooth carburation

retaining screw and lift away the assembly until the nut can be slackened to release the low tension lead terminal. Note carefully the arrangement of the insulating washers on the small bolt which secures the moving contact spring blade and the lead terminal to the fixed contact.

If the contact faces are badly burnt or pitted, or if the moving contact fibre heel shows signs of wear or damage, renew the assembly. It is essential that the points are in good condition if the ignition timing is to be correct; use only genuine Yamaha parts when renewing. If the faces are only mildly marked, clean them using an oilstone or fine emery but be careful to keep them square. If it is necessary to separate the moving contact from the fixed, carefully remove the circlip fitted to the pivot post and note carefully the arrangement of washers at both the pivot post and spring blade fixing. On reassembly, note that the moving contact must be able to move freely; apply a smear of grease to the pivot post. Note also that the low tension lead terminal and the moving contact spring blade must be connected to each other via the small bolt, but that both must be completely insulated from the fixed contact. **The engine will not run if a short-circuit occurs at this point.**

Refit the points to the stator plate and the rotor to the crankshaft. Tighten the rotor retaining nut to the recommended torque setting as

Generator rotor holding tool can be fabricated easily and will prove useful – see Chapter 1

described in Chapter 1, then apply a few drops of oil to the cam lubricating wick.

Certain items of specialised equipment are needed to check the ignition timing, these being rather expensive and infrequently required. It is therefore recommended that the machine be taken to a Yamaha Service Agent for the ignition timing to be checked. The items required are: a dial gauge set (Yamaha Part Number 90890-01173), comprising an accurate dial gauge, a suitable extension rod with a ball-point tip and an adaptor suitable for a 14 mm spark plug thread, and a self-powered points checker (Yamaha Part Number 90890-03031). This last item can be replaced by a proprietary multimeter set to the x 1 ohm resistance scale or by a battery and bulb test circuit. For those who have the necessary equipment, proceed as follows.

Remove the spark plug and the flywheel generator cover. Remove the left-hand side panel and disconnect the black or black/white wire which leads from the points to the ignition switch and the HT coil. Fit the dial gauge adaptor in the spark plug thread and tighten it securely, then screw the extension rod into the gauge, and insert the gauge assembly into the adaptor, securing it with the grub screw. Rotate the flywheel until the piston is at top dead centre (TDC); as the piston approaches TDC the gauge reading will decrease, stop momentarily as TDC is reached, then increase again as the piston descends. Set the gauge to zero as TDC is reached, then rock the flywheel to and fro to make sure that the needle does not go past zero.

The exact time at which the contact breaker points open is determined by the use of a points checker or a multimeter; the gauge needle will swing from 'Closed' to 'Open' (points checker) or will flicker to indicate increased resistance (multimeter). A battery and bulb test circuit can be used so that the bulb lights when the points are closed; note, however, that the bulb will not go out, but will merely glow dimmer as the points open. To make this more obvious to the eye, a high-wattage bulb must be used.

Connect the meter positive (+) terminal to the wire leading from the points and the meter negative (−) terminal to a good earth point on the engine. If a battery and bulb are used, obtain three lengths of wire, connect the battery negative (−) terminal to a good earth point, the battery positive (+) terminal to the bulb contact and the bulb body to the wire leading from the points.

Rotate the flywheel clockwise until a reading of 2 − 3 mm (0.08 − 0.12 in) is shown on the gauge, then rotate it slowly anticlockwise until the points open. The reading shown on the gauge should be precisely 1.80 mm (0.071 in) although a tolerance of 0.15 mm (0.006 in) is allowed on either side of the set figure and any reading within these limits is permissible.

The setting is adjusted by opening or closing the contact breaker points gap to advance or retard respectively the ignition timing. Repeat the procedure to check that the timing is now correct.

When the timing is found to be correct, measure very carefully the points gap. If it is found to be outside the permitted tolerance of 0.30

Contact breaker points must be in good condition for correct ignition timing – renew if worn

Always renew gasket and check that breather hose is unblocked and correctly routed when refitting generator cover

– 0.40 mm (0.012 – 0.016 in) the contact breaker points are excessively worn, either on the contact faces or on the heel of the moving contact, and must be renewed. It is essential that both the contact breaker points gap and the ignition timing setting are kept within the limits specified for each.

Working as described above, fit a new set of contact breaker points; note that it is essential that only genuine Yamaha points should be used. Refit the flywheel and set the points gap to exactly 0.35 mm (0.014 in), then repeat the procedure given above. The ignition timing should be correct, or at least within tolerances.

Note that while timing marks are provided (in the form of a line scribed on the rotor to align with a timing plate mounted either on the condenser or on one of the source coil poles) which apparently make it possible to check the ignition timing using a strobe timing lamp when the engine is running, these marks should not be used unless they have been checked, and found to be accurate, using a dial gauge as described.

3 Check the spark plug

The spark plug supplied as original equipment will prove satisfactory in most operating conditions; alternatives are available to allow for varying altitudes, climatic conditions and the use to which the machine is put. If the spark plug is suspected of being faulty it can be tested only by the substitution of a brand new (not second-hand) plug of the correct make, type, and heat range; always carry a spare on the machine.

Note that the advice of a competent Yamaha Service Agent or similar expert should be sought before the plug heat range is altered from standard. The use of too cold, or hard, a grade of plug will result in fouling and the use of too hot, or soft, a grade of plug will result in engine damage due to the excess heat being generated. If the correct grade of plug is fitted, however, it will be possible to use the condition of the spark plug electrodes to diagnose a fault in the engine or to decide whether the engine is operating efficiently or not. The accompanying series of colour photographs will show this clearly.

It is advisable to carry a new spare spark plug on the machine, having first set the electrodes to the correct gap. Whilst spark plugs do not often fail, a new replacement is well worth having if a breakdown does occur. Ensure that the spare is of the correct heat range and type.

The electrode gap can be assessed using feeler gauges. If necessary, alter the gap by bending the outer electrode, preferably using a proper electrode tool. **Never** bend the centre electrode, otherwise the porcelain insulator will crack, and may cause damage to the engine if particles break away whilst the engine is running. If the outer electrode is seriously eroded as shown in the photographs, or if the spark plug is heavily fouled, it should be renewed. Renew the spark plug at least once annually, regardless of its apparent condition, as it will have passed peak efficiency. Clean the electrodes using a wire brush or a sharp-pointed knife, followed by rubbing a strip of fine emery across the electrodes. If a sand-blaster is used, check carefully that there are no particles of sand trapped inside the plug body to fall into the engine at a later date. For this reason such cleaning methods are no longer recommended; if the plug is so heavily fouled it should be renewed.

Before refitting a spark plug into the cylinder head; coat the threads sparingly with a graphited grease to aid future removal. Use the correct size spanner when tightening the plug, otherwise the spanner may slip and damage the ceramic insulator. The plug should be tightened by hand only at first and then secured with a quarter turn of the spanner so that it seats firmly on its sealing ring. If a torque wrench is available, tighten the plug to the specified torque setting.

Never overtighten a spark plug otherwise there is risk of stripping the threads from the cylinder head, especially as it is cast in light alloy. A stripped thread can be repaired without having to scrap the cylinder head by using a 'Helicoil' thread insert. This is a low-cost service, operated by a number of dealers.

4 Clean the fuel tap

Switch the tap to the 'Off' position and unscrew the filter bowl from the tap base. Check the condition of the sealing O-ring and renew it if it is seriously compressed, distorted or damaged. Thoroughly clean the filter bowl; if signs of dirt or water are found in the petrol, the tap should be removed from the tank as described in Chapter 2 so that the tap filter can be cleaned and the tank flushed out.

On reassembly, tighten the bowl by just enough to nip up the O-ring; do not overtighten it as this will distort the O-ring and promote the risk of fuel leakage. If leaks are found they can be cured only by the renewal of the defective seal.

5 Oil the control cables and grease the twistgrip, stand pivot and controls

At regular intervals the footrests, prop stand, the rear brake pedal and the throttle twistgrip, must be dismantled so that all traces of corrosion and dirt can be removed, and the various components greased. This operation must be carried out to prevent excessive wear and to ensure that the various components can be operated smoothly and easily, in the interests of safety. The opportunity should be taken to examine closely each component, renewing any that show signs of excessive wear or of any damage.

The twistgrip is removed by unscrewing the screws which fasten both halves of the handlebar right-hand switch assembly. The throttle cable upper end nipple can be detached from the twistgrip with a suitable pair of pliers and the twistgrip slid off the handlebar end. Carefully clean and examine the handlebar end, the internal surface of the twistgrip, and the two halves of the switch cluster. Remove any rough burrs with a fine file, and apply a coating of grease to all the bearing surfaces. Slide the twistgrip back over the handlebar end, insert the throttle cable end nipple into the twistgrip flange, and reassemble the switch cluster. Check that the twistgrip rotates easily and that the throttle snaps shuts as soon as it is released. Tighten the switch retaining screws securely, but do not overtighten them.

Although the regular daily checks will ensure that the control cables are lubricated and maintained in good order, it is recommended that a positive check is made on each cable at this mileage/time interval to ensure that any faults will not develop unnoticed to the point where smooth and safe control operation is impaired. If any doubt exists about the condition of any of the cables, the component in question should be removed from the machine for close examination. Check the outer cables for signs of damage, then examine the exposed portions of the inner cables. Any signs of kinking or fraying will indicate that renewal is required. To obtain maximum life and reliability from the cables they should be thoroughly lubricated using light machine oil. To do the job properly and quickly use one of the hydraulic cable oilers available from most motorcycle shops. Free one end of the cable and assemble the cable oiler as described by the manufacturer's instructions. Operate the oiler until oil emerges from the lower end, indicating that the cable is lubricated throughout its length. This process will expel any dirt or moisture and will prevent its subsequent ingress.

If a cable oiler is not available, an alternative is to remove the cable from the machine. Hang the cable upright and make up a small funnel arrangement using plasticine or by taping a plastic bag around the upper end. Fill the funnel with oil and leave it overnight to drain

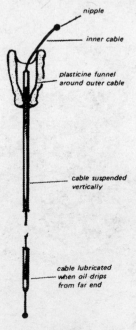

Oiling a control cable

through. Note that where nylon-lined cables are fitted, they should be used dry or lubricated with a silicone-based lubricant suitable for this application. On no account use ordinary engine oil because this will cause the liner to swell, pinching the cable. The throttle/oil pump cable junction box should be dismantled, cleaned and reassembled using a 'dry' graphite-based lubricant or any lightweight grease or aerosol spray lubricant. Do not use ordinary grease as this will at best make the throttle action heavy; at worst it could cause an accident by preventing the smooth return of the throttle.

Check all pivots and control levers, cleaning and lubricating them to prevent wear or corrosion. Where necessary, dismantle and clean any moving part which may have become stiff in operation.

When refitting the cables onto the machine, ensure that they are routed in easy curves and that full use is made of any guide or clamps that have been provided to secure the cable out of harm's way. Adjustment of the individual cables is described under Routine Maintenance tasks.

Be very careful to ensure that all controls are correctly adjusted and are functioning correctly before taking the machine out on the road.

6 Check the carburettor, throttle cable and fuel pipe

If rough running of the engine has developed, some adjustment of the carburettor pilot setting and tick over speed may be required. If this is the case refer to Chapter 2 for details. Do not make these adjustments unless they are obviously required; there is little to be gained by unwarranted attention to the carburettor. Complete carburettor maintenance by removing the drain plug on the float chamber, turning the petrol on, and allowing a small amount of fuel to drain through, thus flushing any water or dirt from the carburettor. Refit the drain plug securely and switch the petrol off.

Once the carburettor has been checked and reset if necessary, the throttle cable free play can be checked. Open and close the throttle several times, allowing it to snap shut under its own pressure. Ensure that it is able to shut off quickly and fully at all handlebar positions, then check that there is 1.0 mm (0.04 in) free play in the cable at the carburettor top. If little or no free-play exists, fully slacken the cable adjuster at or just below the twistgrip and check that the throttle cable is not trapped at any point. Use the adjuster on the carburettor top to set the correct amount of free play, tighten its locknut and replace the rubber cover. Free play in the twistgrip/junction box section of throttle cable is set using the adjuster at the twistgrip. On TY50 and TY80 models there should be 0.5 – 1.0 mm (0.02 – 0.04 in) of free play in the cable itself. On TY125 and TY175 models free play is measured in terms of twistgrip rotation, ie there should be 3 – 6 mm (0.12 – 0.24 in) of free twistgrip rotation before the cable is pulled. Open and close the throttle again to settle the cable and to check that adjustment is not disturbed. Do not forget to check the oil pump cable.

Give the pipe which connects the fuel tap and carburettor a close visual examination, checking for cracks or any signs of leakage. In time, the synthetic rubber pipe will tend to deteriorate, and will eventually leak. Apart from the obvious fire risk, the leaking fuel will affect fuel economy. If the pipe is to be renewed, always use the correct replacement type to ensure a good leak-proof fit. Never use natural tubing because this will tend to break up when in contact with petrol and will obstruct the carburettor jets.

7 Check the oil pump settings

The oil pump adjustments are in two parts: checking the pump minimum stroke setting to ensure that the pump is delivering the correct amount of oil and checking the oil pump cable adjustment to ensure that the pump is correctly synchronised with the carburettor. Both tasks must commence with the removal of the oil pump cover from the front of the crankcase right-hand cover.

Checking and resetting the pump minimum stroke

The pump's minimum stroke adjustment should be checked first. Start the engine and allow it to idle. Observe the front end of the pump unit, where it will be noticed that the pump adjustment plate moves in

Regularly lubricate all pivots and controls – dismantle for cleaning and greasing where possible

Carburettor/junction box section of throttle cable is adjusted at carburettor top ...

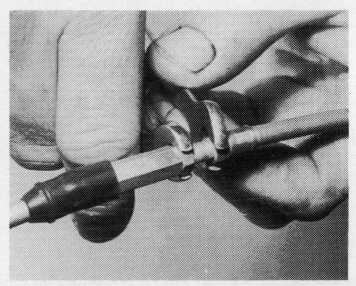

... while adjuster at twistgrip is used to adjust upper cable

and out. When the plate is out to its fullest extent, stop the engine and measure the gap between the plate and the raised boss of the pump pulley using feeler gauges. Do not force the feeler gauge into the gap – it should be a light sliding fit. Make a note of the reading, then repeat the procedure several times. The largest gap is indicative that the pump is at its minimum stroke position. If the pump is set up correctly, the gap should be 0.20 – 0.25 mm (0.008 – 0.010 in) on all models except the TY80, where it should be 0.30 – 0.35 mm (0.012 – 0.014 in).

If the pump setting is found to be incorrect, slacken and remove the adjustment plate retaining nut, then withdraw the spring washer and the adjustment plate. The pump stroke is set by adding or removing shims from behind the adjustment plate, these shims being available from Yamaha dealers in thicknesses of 0.3, 0.5 and 1.0 mm (0.0118, 0.0197 and 0.0394 in). When the shim thickness is correct, refit the adjustment plate, the spring washer, and the retaining nut, then start the engine and recheck the minimum stroke setting. If necessary, repeat the procedure to ensure that the setting is correct.

Note that early models have a white nylon pinion with a knurled rim fitted to the rear of the pump unit; the purpose of this was to provide a means of operating the pump independently of the engine during both this procedure and the bleeding procedure described in Chapter 2. Its use was, however, found to be both time-consuming and unnecessary and it was deleted on later models; the procedure given above is suitable for all models. Owners of early machines may rotate the pinion (clockwise, viewed from the rear) or start the engine, as desired, to set the pump to its minimum stroke position.

Checking the oil pump cable adjustment

Set the pump to its minimum stroke position as described above then carefully rotate the twistgrip until all slack has **just** been removed from the throttle/oil pump cable. At this point the pulley guide pin should align exactly with the mark on the pulley, as shown in the accompanying illustrations. On TY50, TY80 and TY175 models the correct mark is the circular one immediately above or below (depending on model and year) the raised rectangular line. On TY125 models the correct mark is in the form of a raised square-shape. If the guide pin and mark do not align, check that the pulley is free to rotate fully home (ie in the pump minimum stroke position), then use the oil pump cable adjuster to bring the two into line. When the setting is correct, tighten the adjuster locknut, replace the rubber sleeve and oil the exposed length of cable inner wire and the pump moving parts before refitting the pump cover. Note that this setting must be checked after any work is carried out on any part of the throttle/oil pump cable.

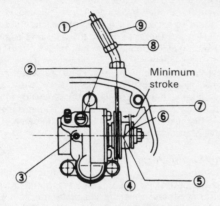

Oil pump minimum stroke setting

1	Pump cable	6	Adjustment plate
2	Bleed screw	7	Pump pulley guide pin
3	Blind plug	8	Locknut
4	Pump pulley	9	Cable adjuster
5	Alignment mark		

Measuring oil pump minimum stroke

Shims of varying thicknesses are used to adjust pump stroke

Pulley guide pin must align with correct mark on pulley, as shown

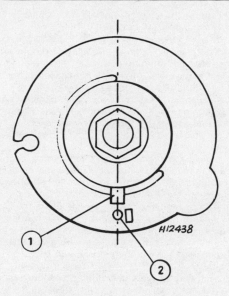

Oil pump cable alignment mark – TY50, 80 and 175

1   *Pulley guide pin*   2   *Alignment mark*

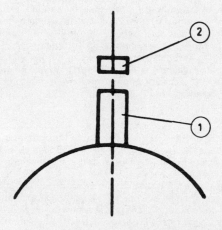

Oil pump cable alignment mark – TY125

1   *Pulley guide pin*   2   *Alignment mark*

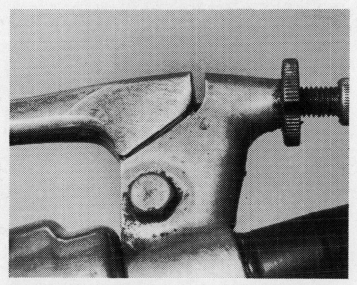

Clutch cable free play is measured between lever end and clamp, as shown ...

8   *Adjust the clutch*

The clutch is adjusted correctly when there is 2 – 3 mm (0.08 – 0.12 in) of free play in the cable, measured between the butt end of the handlebar lever and its clamp, and when the clutch is operating correctly with no signs of slip or drag.

Normal adjustment is made at the cable lower (TY50, TY80) or mid-way (TY125, TY175) adjuster, reserving the lever clamp adjuster for quick roadside adjustments. If this is no longer possible, or if there are signs of slip or drag, the mechanism must be adjusted as follows.

TY50, TY80

Slacken their locknuts and screw fully in both adjusters to gain the maximum cable free play. Remove the cover or cap from the crankcase left-hand cover to expose the clutch adjuster. Slacken the adjuster locknut and unscrew (anti-clockwise) the adjuster screw to ensure that there is no pressure on it, then screw it in (clockwise) until it seats lightly; do not overtighten it as this is the point where the mechanism is starting to lift the clutch pressure plate. Set the required free play by unscrewing (anti-clockwise) the adjusting screw through $\frac{1}{4}$ turn, then hold it while the locknut is tightened securely. Replace the cap or cover and use the cable lower adjuster to set the specified free play at the handlebar. Apply a few drops of oil to the lever pivot, to the adjuster threads and to all exposed lengths of inner cable.

Use adjuster shown to adjust oil pump control cable

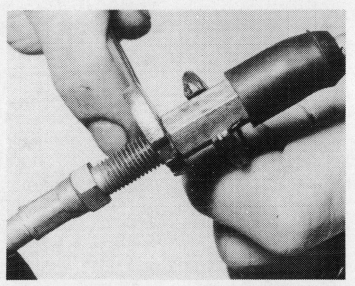

... and is adjusted using cable adjusters – TY125 and 175

TY125, TY175

Slacken both cable adjuster locknuts and remove the crankcase left-hand cover, then slacken the mechanism adjuster locknut. Screw in the adjusting screw until it seats fully; do not overtighten the screw as it engages on the release shaft via an eccentric pin which may be sheared off. (Note that the clutch release shaft/operating lever assembly will move up and down as the screw is rotated). Screw fully in the cable adjuster at the handlebar clamp, then screw in the mid-way adjuster until the operating lever returns under spring pressure to a point slightly behind the axis of the input shaft, ie directly across the engine/gearbox unit. Unscrew (anti-clockwise) the adjusting screw until light pressure indicates that the release shaft has risen into contact with the clutch pushrod, removing all free play. Set the specified free play by screwing in the adjuster screw $\frac{1}{8}$ turn and tighten securely the locknut. Use the mid-way adjuster first, followed by the handlebar adjuster, to set the specified cable free play at the handlebar then tighten both adjuster locknuts and replace the rubber covers. Apply a few drops of oil to the lever pivots, to the adjuster threads and to all exposed lengths of inner cable, then refit the crankcase cover.

9 Change the transmission oil

Oil changes are much quicker and more efficient if the machine is taken for a journey long enough to warm the oil up to normal operating temperature so that it is thin and is holding any impurities in suspension. Position a container of at least 700 cc (1.2 Imp pint) capacity beneath the engine. Remove the drain plug from the underside of the engine unit. Whilst waiting for the oil to drain, examine the plug sealing washer and renew if damaged.

On completion of draining, refit the plug and tighten it to the recommended torque setting. Remove the filler plug adjacent to the kickstart lever and replenish the gearbox with the specified quantity of SAE 10W/30 oil, see Chapter 2. Check the oil level as described under the six-weekly/1000 mile heading and refit the filler plug.

10 Check the battery

It is essential that the battery be maintained in excellent condition to prolong its life. In addition to the check of the electrolyte level, the condition of the terminals should be examined. Clean away all traces of dirt and corrosion, scraping the terminals and connections with a knife and using emery cloth to finish off. Remake the connections while the joint is still clean and then smear the assembly with petroleum jelly (not grease) to prevent recurrence of the corrosion. Finish off by checking that the battery is securely clamped in its mountings and that the vent tube is quite clean and free from kinks or blockages.

The electrolyte level must be maintained between the level marks on the casing. Top up, if necessary, using distilled water.

11 Check the suspension and steering head bearings

Place the machine on a stand so that the front wheel is clear of the ground. If necessary, place blocks below the crankcase to prevent the motorcycle from tipping forwards.

Grasp the front fork legs near the wheel spindle and push and pull firmly in a fore and aft direction. If play is evident between the top and bottom fork yokes and the steering head, the steering head bearings are in need of adjustment. Imprecise handling or a tendency for the front forks to judder may be caused by this fault.

Bearing adjustment is correct when the adjuster ring is tightened, until resistance to movement is felt and then loosened $\frac{1}{8}$ to $\frac{1}{4}$ of a turn. The adjuster ring should be rotated by means of a C-spanner after slackening the steering stem top bolt (fork crown bolt).

Take great care not to overtighten the adjuster ring. It is possible to place a pressure of several tons on the head bearings by over-tightening even though the handlebars may seem to turn quite freely. Overtight bearings will cause the machine to roll at low speeds and give imprecise steering. Adjustment is correct if there is no play in the bearings and the handlebars swing to full lock either side when the machine is on the centre stand with the front wheel clear of the ground. Only a light tap on each end should cause the handlebars to swing. Secure the adjuster ring by tightening the steering stem top bolt to the specified torque setting, then check that the setting has not altered.

At the same time as the steering head bearings are checked, take the opportunity to examine closely the front and rear suspension. Ensure that the front forks work smoothly and progressively by pumping them up and down whilst the front brake is held on. Any faults revealed by this check should be investigated further, as any deterioration in the stability of the machine can have serious conse-quences. Check carefully for signs of leaks around the front fork oil seals. If any damage is found, it must be repaired immediately as described in the relevant Sections of Chapter 4. Examine the rear suspension in the same way and check for wear in the swinging arm pivot by pushing and pulling horizontally at its rear end. There should be no discernible play at the pivot. Where a grease nipple is provided, pump grease into the pivot bearing until fresh grease is seen at both ends of the swinging arm.

12 Check the wheels and tyres

Raise clear of the ground the wheel to be examined, using blocks positioned beneath the engine. Check the wheel spins freely; if necessary, slacken the brake adjuster and in the case of the rear wheel, detach the final drive chain.

Examine the rim for serious corrosion or impact damage. Slight deformities can often be corrected by adjusting spoke tension. Serious damage and corrosion will necessitate renewal, which is best left to an expert. A light alloy rim will prove more corrosion resistant.

Place a wire pointer close to the rim and rotate the wheel to check it for runout. If the rim is more than 2.0 mm (0.08 in) out of true in the radial or axial planes, check spoke tension by tapping them with a screwdriver. A loose spoke will sound quite different to those around

Adjusting clutch release mechanism – TY125 and 175

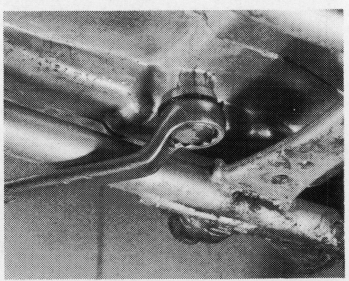

Remove drain plug to drain transmission oil – crankcase bashplate removed for clarity

it. Worn bearings will also cause rim runout.

Adjust spoke tension by turning the square-headed nipples with the appropriate spoke key which can be purchased from a dealer. With the spokes evenly tensioned, remaining distortion can be pulled out by tightening the spokes on one side of the wheel and slackening those directly opposite. This will pull the rim across whilst maintaining spoke tension.

More than slight adjustment will cause the spoke ends to protrude through the nipple and chafe the inner tube, causing a puncture. Remove the tyre and tube and file off the protruding ends. The rim band protects the tube against chafing, check it is in good condition before fitting.

Check spoke tension and general wheel condition regularly. Frequent cleaning will help prevent corrosion. Replace a spoke immediately because the load taken by it will be transferred to adjacent spokes which may fail in turn.

An out of balance wheel will produce a hammering effect through the steering at high speed. Spin the wheel several times. A well balanced wheel will come to rest in any position. One that comes to rest in the same position will have its heaviest part downward and weights must be added to a point diametrically opposite until balance is achieved. Where the tyre has a balance mark on its sidewall (usually a coloured spot), check it is in line with the valve.

To check the wheel bearings, grasp each wheel firmly at the top and bottom and attempt to rock it from side to side; any free play indicates worn bearings which must be renewed as described in Chapter 5. Make a careful check of the tyres, looking for signs of damage to the tread or sidewalls and removing any embedded stones etc. Renew the tyre if the tread is excessively worn or if it is damaged in any way.

13 Check all fittings, fasteners, lights and signals
Check around the machine, looking for loose nuts, bolts or screws, retightening them as necessary. Check the stand and lever pivots or security and lubricate them with light machine oil or engine oil. Make sure that the stand springs are in good condition.

It is advisable to lubricate the handlebar switches and stop lamp switches with WD40 or a similar water dispersant lubricant. This will keep the switches working properly and prolong their life, especially if the machine is used in adverse weather conditions. Check that all lights, turn signals, the horn and speedometer are working properly and that their mountings and connections are securely fastened.

Six monthly, or every 4000 miles (6000 km)

Complete the operations listed under the previous mileage/time headings, then carry out the following:

1 Overhaul the carburettor
If this has not been done previously, the carburettor should be removed from the machine, to be dismantled and cleaned thoroughly. On reassembly, check the settings, as described in Chapter 2.

2 Grease the rear suspension pivot
Remove the rear wheel as described in Chapter 5 and withdraw the swinging arm as described in Chapter 4. Clean all components and check them for wear or damage, renewing any component as necessary. On reassembly, pack the pivot bearings with grease (where applicable).

3 Change the fork oil
On machines not fitted with front fork drain plugs, each fork leg must be removed from the yokes and inverted to pump out the old oil. Refer to Chapter 4. On machines with drain plugs, place a sheet of cardboard against the wheel to keep oil off the brake or tyre, place a suitable container under the fork leg and remove the drain plug. Pump the forks several times to expel as much oil as possible, then repeat the process on the remaining leg. Leave the machine for a few minutes to allow any residual oil to drain to the bottom, then pump the forks again to remove it.

Renewing their sealing washers if worn or damaged, refit and tighten securely the drain plugs, then remove the fork leg top bolts. Add to each leg exactly the correct amount of the specified grade of fork oil (see Chapter 4 Specifications) before refitting the top bolts. Tighten the top bolts to the specified torque setting and check that the forks move smoothly and correctly throughout their full travel.

Note that the fork action can be varied to suit the rider's needs by using different grades of oil; the lighter the oil, the lighter the damping effect. Proprietary fork oils are available in several grades from most good motorcycle dealers.

Annually, or every 8000 miles (12 000 km)

Every year the machine should be taken out of service for a major overhaul in which all previously listed maintenance operations should be carried out in full and with care. If they have not been carried out already during the course of maintenance or during repair work, the following three operations should be undertaken at this interval:

1 Grease the steering head bearings
Referring to the relevant Sections of Chapter 4, dismantle the front forks and steering head, clean all components and check them for wear, renewing any worn items. Reassemble, packing the steering head bearings with fresh grease.

2 Overhaul the brakes
If this task was not carried out at the six-weekly/1000 mile interval, the wheels must be removed so that the brakes can be dismantled for thorough cleaning, checking, and lubrication. Refer to Chapter 5.

3 Grease the wheel bearings and speedometer drive
This would fit in conveniently with the previous operation. These components must be checked for wear and packed with fresh grease at least once annually. Refer to Chapter 5.

Chapter 1 Engine, clutch and gearbox

Contents

Specifications

Note: unless otherwise stated, specifications are the same for all models

Engine

Type	Single cylinder, air-cooled, two-stroke			
Dimensions:	**TY50**	**TY80**	**TY125**	**TY175**
Capacity	49cc (2.99 cu in)	72cc (4.39 cu in)	123cc (7.50 cu in)	171cc (10.43 cu in)
Bore	40 mm (1.58 in)	47 mm (1.85 in)	56 mm (2.21 in)	66 mm (2.60 in)
Stroke	39.7 mm (1.56 in)	42 mm (1.65 in)	50 mm (1.97 in)	50 mm (1.97 in)
Compression ratio	6.8 : 1	7.0 : 1	7.1 : 1	6.8 : 1
Output:				
Maximum power – bhp @ rpm	2.9 @ 5500	4.9 @ 6000	N/Av	11.4 @ 7000
Maximum torque – lbf ft @ rpm	2.9 @ 5000	4.4 @ 5500	N/Av	10.1 @ 7000

Cylinder head

Gasket face maximum warpage	0.02 mm (0.0008 in)
Combustion chamber volume:	
TY50	5.1 $\pm$ 0.15cc (0.1795 $\pm$ 0.0053 fl oz)
All other models	N/Av
Gasket thickness:	
TY50	1.2 mm (0.0472 in)
All other models	N/Av

Compression pressure – engine warm:
    TY80 .... 5.5 kg/cm^2 (78 psi) at cranking speed
    All other models .... N/Av

Cylinder barrel
Maximum taper .... 0.050 mm (0.0020 in)
Maximum ovality:
    Except TY80 .... 0.010 mm (0.0004 in)
    TY80 .... 0.005 mm (0.0002 in)
Piston/cylinder standard clearance:
    TY50 .... 0.030 – 0.035 mm (0.0012 – 0.0014 in)
    TY80, TY125 .... 0.035 – 0.040 mm (0.0014 – 0.0016 in)
    TY175 .... 0.040 – 0.045 mm (0.0016 – 0.0018 in)
Piston/cylinder maximum clearance .... 0.100 mm (0.0039 in)

Piston and piston rings
Piston standard OD:
    TY50 .... 39.96 – 40.00 mm (1.5732 – 1.5748 in)
    TY80 .... 46.96 – 47.00 mm (1.8488 – 1.8504 in)
    TY125 .... 55.97 – 56.00 mm (2.2035 – 2.2047 in)
    TY175 .... 65.96 – 66.00 mm (2.5969 – 2.5984 in)
Oversizes available:
    Except TY80 .... 0.25, 0.50, 0.75 and 1.00 mm (0.010, 0.020, 0.030 and 0.040 in)
    TY80 .... 0.25, 0.50 mm (0.010, 0.020 in)
Top piston ring type:
    TY50, TY80 .... Keystone
    TY125, TY175 .... Dykes
Second piston ring type .... Plain, with expander
Ring dimensions – note information available for TY125 only:
    Top ring standard thickness .... 2.5 mm (0.0984 in)
    Top ring standard width .... 2.5 mm (0.0984 in)
    Second ring standard thickness .... 1.5 mm (0.0591 in)
    Second ring standard width .... 1.8 mm (0.0709 in)
Ring/piston groove clearance – note, information
not applicable to TY125/175 top (Dykes) ring: ...... 0.03 – 0.08 mm (0.0012 – 0.0032 in)
Ring end gap – installed:
    Except TY175 .... 0.15 – 0.35 mm (0.0059 – 0.0138 in)
    TY175 .... 0.30 – 0.50 mm (0.0118 – 0.0197 in)

Ring end gap-free:	TY50	TY80	TY125	TY175
Top	5.0 mm (0.1969 in)	7.5 mm (0.2953 in)	5.5 mm (0.2165 in)	8.5 mm (0.3347 in)
Second	6.0 mm (0.2362 in)	4.0 mm (0.1575 in)	5.5 mm (0.2165 in)	4.5 mm (0.1772 in)

Crankshaft
Small-end bearing maximum radial play .... 0.05 mm (0.0020 in)
Big-end bearing deflection – at small-end .... 0.80 – 1.00 mm (0.0315 – 0.0394 in)
Service limit .... 2.00 mm (0.0787 in)
Big-end side clearance:
    TY50 .... 0.20 – 0.50 mm (0.0079 – 0.0197 in)
    Service limit .... 0.60 mm (0.0236 in)
    TY80 .... 0.20 – 0.40 mm (0.0079 – 0.0158 in)
    Service limit .... 0.50 mm (0.0197 in)
    TY125 .... 0.25 – 0.75 mm (0.0098 – 0.0295 in)
    Service limit .... 0.80 mm (0.0315 in)
    TY175 .... 0.20 – 0.40 mm (0.0079 – 0.0158 in)
    Service limit .... 0.60 mm (0.0236 in)
Maximum runout .... 0.03 mm (0.0012 in)
Width across flywheels:
    TY50, TY80 .... 37.90 – 37.95 mm (1.4921 – 1.4941 in)
    TY125, TY175 .... 55.90 – 55.95 mm (2.2008 – 2.2028 in)

Primary drive
Type .... Gear
Reduction ratio:
    TY50, TY80 .... 3.578 : 1 (68/19T)
    TY125, TY175 .... 3.895 : 1 (74/19T)

Clutch
Type .... Wet, multi plate

	TY50	TY80	TY125, TY175
Number of friction plates	3 (2, later models)	3	5
Number of plain plates	2 (1, later models)	2	4
Number of springs	4	4	5

Friction plate thickness:
    TY50, TY80 .... 3.5 mm (0.1378 in)
    Service limit .... 3.2 mm (0.1260 in)
    TY125, TY175 .... 3.0 mm (0.1181 in)
    Service limit .... 2.7 mm (0.1063 in)
Plain plate thickness:
    TY50, TY80 .... 1.6 mm (0.0630 in)
    TY125, TY175 .... 1.2 mm (0.0472 in)
Plain plate maximum warpage .... 0.05 mm (0.0020 in)
Spring standard free length .... 31.5 mm (1.2402 in)
Service limit .... 30.5 mm (1.2008 in)
Maximum difference between spring free lengths .... 1.0 mm (0.0394 in)
Push rod maximum warpage .... 0.5 mm (0.0197 in)
Outer drum axial play:
    TY50, TY125, TY175 .... 0.07 – 0.10 mm (0.0028 – 0.0039 in)
    Service limit .... 0.15 mm (0.0059 in)
    TY80 .... 0.10 – 0.15 mm (0.0039 – 0.0059 in)
    Service limit .... 0.30 mm (0.0118 in)

	TY50, TY80	**TY125, TY175**
Outer drum centre bush ID	N/Av	22.995 – 23.016 mm (0.9053 – 0.9061 in)
Bearing sleeve OD	N/Av	22.967 – 22.980 mm (0.9042 – 0.9047 in)
Bush/sleeve clearance	N/Av	0.015 – 0.049 mm (0.0006 – 0.0019 in)
Input shaft OD	N/Av	16.930 – 16.945 mm (0.6665 – 0.6671 in)
Bearing sleeve ID	N/Av	16.994 – 17.012 mm (0.6691 – 0.6698 in)
Shaft/sleeve clearance	N/Av	0.049 – 0.082 mm (0.0019 – 0.0032 in)

Gearbox

	TY50	**TY80**	**TY125, TY175**
Number of ratios	5-speed	4-speed	6-speed
Reduction ratios:			
1st	3.250 : 1 (39/12T)	3.250 : 1 (39/12T)	3.091 : 1 (34/11T)
2nd	2.000 : 1 (34/17T)	2.000 : 1 (34/17T)	2.462 : 1 (32/13T)
3rd	1.428 : 1 (30/21T)	1.428 : 1 (30/21T)	1.875 : 1 (30/16T)
4th	1.125 : 1 (27/24T)	1.125 : 1 (27/24T)	1.421 : 1 (27/19T)
5th	0.962 : 1 (25/26T)	N/App	1.000 : 1 (23/23T)
6th	N/App	N/App	0.769 : 1 (20/26T)
Gearbox shafts maximum runout	0.015 mm (0.0006 in)		
Kickstart friction clip resistance	0.8 – 1.5 kg (1.76 – 3.31 lb)		

Final drive

Type .... Chain and sprockets

	TY50P	**TY50M – early models**	**TY50M – later models**
Reduction ratio	3.000 : 1 (42/14T)	3.429 : 1 (48/14T)	4.182 : 1 (46/11T)
Chain size	420 ($\frac{1}{2}$ x $\frac{1}{4}$ in) x 104 lks	420 ($\frac{1}{2}$ x $\frac{1}{4}$ in) x 104 lks	420 ($\frac{1}{2}$ x $\frac{1}{4}$ in) x 104 lks
	TY80	**TY125**	**TY175**
Reduction ratio	3.417 : 1 (41/12T)	2.176 : 1 (37/17T)	3.923 : 1 (51/13T)
Chain size	420 ($\frac{1}{2}$ x $\frac{1}{4}$ in) x 106 lks	428 ($\frac{1}{2}$ x $\frac{5}{16}$ in) x 106 lks	428 ($\frac{1}{2}$ x $\frac{5}{16}$ in) x 114 lks

Torque wrench settings

Component	kgf m	lbf ft
Spark plug:		
TY50, TY80	2.5 - 3.0	18.0 - 22.0
TY125, TY175	2.0	14.5
Cylinder head nuts:		
TY50, TY80	0.8 - 1.0	6.0 - 7.0
TY125, TY175	2.1 - 2.5	15.0 - 18.0
Crankcase and cover fastening screws	0.9 - 1.1	6.5 - 8.0
Transmission oil drain plug:		
TY50, TY80	3.5 - 4.0	25.0 - 29.0
TY125, TY175	2.0 - 2.5	14.5 - 18.0
Clutch centre nut:		
TY50, TY80	4.0 - 4.5	29.0 - 32.5
TY125, TY175	6.0 - 7.0	43.0 - 50.5
Clutch spring bolts	0.7 - 1.0	5.0 - 7.0
Primary drive gear nut:		
TY50, TY80	4.0 - 4.5	29.0 - 32.5
TY125, TY175	7.0	50.5
Generator rotor nut:		
TY50	5.0 - 7.0	36.0 - 50.5
TY80	3.5 - 4.0	25.0 - 29.0
TY125, TY175	6.0 - 7.0	43.0 - 50.5
Gearbox sprocket nut:		
TY50, TY80	4.0 - 4.5	29.0 - 32.5
TY125, TY175	6.0 - 7.0	43.0 - 50.5

Engine mounting bolts – front, rear upper:			
TY50, TY80	2.2 - 3.0		16.0 - 22.0
TY125, TY175	1.0		7.0
Engine rear lower mounting bolt	3.0 - 4.0		22.0 - 29.0

1 General description

The crankshaft is a built-up assembly which rotates on two ball journal main bearings and is fitted with needle roller bearings at the connecting rod big- and small-ends. A flywheel generator is fitted on its left-hand end, a gear on its right-hand end providing the drive for the clutch and oil pump. The clutch is a wet multi-plate type mounted on the right-hand end of the gearbox input shaft, the gearbox itself being of the constant-mesh type built in unit with the engine. All engine and gearbox components are housed in the vertically split, aluminium alloy engine unit castings.

The engine/gearbox unit is simple in design and construction, requiring the bare minimum of special tools during dismantling and overhaul, and has proved itself to be extremely reliable in service, with no major faults.

2 Operations with the engine/gearbox unit in the frame

The following items can be removed with the engine/gearbox unit in the frame:

a) Cylinder head, barrel and piston, carburettor and reed valve
b) Oil pump
c) Clutch assembly and primary drive gear pinion
d) Gear selector mechanism external components
e) Kickstart mechanism
f) Flywheel generator assembly
g) Gearbox sprocket and, where fitted, neutral switch

3 Operations with the engine/gearbox unit removed from the frame

It will be necessary to remove the complete engine/gearbox unit from the frame and to separate the crankcase halves to gain access to the following components:

a) Crankshaft, main bearings and oil seals
b) Gearbox shafts and bearings, gear selector drum and forks

Note that while it is possible to remove and refit the crankshaft main bearing oil seals without separating the crankcase halves, this requires a great deal of care and is not recommended for the reasons given in Section 14.

4 Removing the engine/gearbox unit from the frame

1 If the machine is dirty, it is advisable to wash it thoroughly before starting any major dismantling work. This will make work much easier and will prevent the risk of disturbed lumps of caked-on dirt falling into some vital component.
2 Drain the transmission oil as described in Routine Maintenance. While the oil is draining remove, where fitted, the left-hand side panel and unlock and raise, or remove completely (as appropriate), the seat. Remove the fuel tank and, where fitted, the crankcase bashplate.
3 Note that whenever any component is removed, all mounting nuts, bolts or screws should be refitted in their original locations with their respective washers and mounting rubbers (if any).
4 Work is made much easier if the machine is lifted to a convenient height on a purpose-built ramp or a platform constructed of planks and concrete blocks. Ensure that the wheels are chocked with wooden blocks so that the machine cannot move and that it is securely tied down so that it cannot fall, also that the side stand is supporting it correctly.
5 On machines so equipped, disconnect the battery to prevent any risk of short circuits. If the machine is to be out of service for some

time, remove the battery and give it regular refresher charges as described in Chapter 6.
6 Except for TY80 models, remove its single mounting bolt and pull the exhaust tailpipe/silencer backwards out of the main exhaust pipe. On TY50/80 models unscrew the exhaust pipe/cylinder barrel retaining ring; on TY125/175 models remove the two exhaust front mounting nuts. On all models, remove the single bolt securing the main exhaust pipe to the frame and withdraw the pipe from the machine.
7 Marking its shaft so that it can be refitted in the same position, remove fully the kickstart lever pinch bolt and pull the lever off the shaft-splines. Remove its mounting screws and withdraw the oil pump cover from the front of the crankcase right-hand cover.
8 Disconnect the oil tank/pump feed pipe at the tank union and plug the pipe to prevent the loss of oil and the entry of dirt or air into the system. Plug the pipe with a screw or bolt of suitable size; the tank union can be blanked off by replacing the feed pipe either with the free end of the tank's breather pipe, thus forming a closed circuit, or with a previously prepared short length of tubing with one end plugged. Release the feed pipe from any clamps or ties securing it to the frame.
9 Using a small screwdriver to ease back its tubular metal retaining clamp, pull the oil pump/engine feed pipe off its union on the carburettor or intake stub, as appropriate, then plug it with a screw or bolt of suitable size. Disconnect the oil pump control cable from the pump pulley, then slacken the locknut at the lower end of the cable adjuster and unscrew the adjuster from the crankcase right-hand cover. It may be necessary to unscrew the adjusting nut to permit the adjuster to be rotated past the cylinder barrel; take care not to damage the cable.
10 Unscrew the carburettor top, withdraw the throttle slide assembly and secure the complete throttle/oil pump cable assembly to the frame top tube, out of harm's way. Wrap the throttle slide and needle in clean rag to prevent damage. Slacken the clamp securing the air filter hose to the carburettor and slacken the clamp, or remove the two bolts, as appropriate, securing the carburettor to the intake stub, then withdraw the carburettor, disconnecting first the air filter hose.
11 Marking its shaft so that it can be refitted in the same position, remove fully the gearchange pedal pinch bolt and pull the lever off the shaft splines. Remove its mounting screws and withdraw the crankcase left-hand cover. On TY50/80 models this will also remove the clutch cable and operating mechanism; the cable need be disconnected only if necessary. On TY125/175 models, slacken their locknuts and screw in fully both cable adjuster nuts, flatten back the small metal tang which prevents the cable end nipple from jumping out of the operating lever end, then disconnect the cable from the operating lever. Either disengage the cable from its bracket or unbolt the bracket from the cylinder barrel, as required.
12 Flatten the raised portion of its locking washer, then remove the gearbox sprocket retaining nut, applying the rear brake hard to prevent rotation. Remove the sprocket from the shaft, disengage it from the chain and hang the chain over the swinging arm pivot, noting that it may be necessary to slacken the chain adjustment to permit the sprocket to be removed.
13 Tracing the generator lead from the crankcase top surface up to the connectors joining it to the main loom, disconnect all electrical wires and release the lead from any clamps or ties securing it to the frame. Pull the spark plug cap off the plug. The engine/gearbox unit should now be retained only by its three mounting bolts; check carefully that all components have been removed or disconnected so that nothing will prevent or hinder the removal of the unit.
14 On TY50/80 models remove their retaining nuts and lock washers and tap out the three mounting bolts, then withdraw the engine/gearbox unit. On TY125/175 models unscrew the lower rear mounting bolt and withdraw it, then remove the retaining nuts of the remaining two bolts and tap out the bolts; note that the front mounting bolt may have a D-shaped washer fitted on each side. Lift the engine/gearbox unit out of the frame. On all models, if any bolt is locked in place with corrosion or dirt, apply a liberal quantity of penetrating fluid, allow time for it to work and then release the bolt by rotating it with a spanner before tapping it out with a hammer and drift.

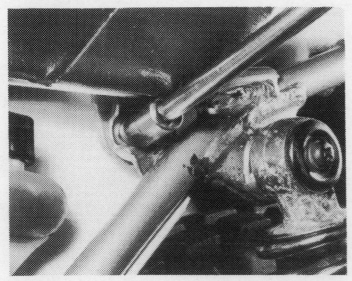
4.2a Withdraw seat mounting bolts and remove seat

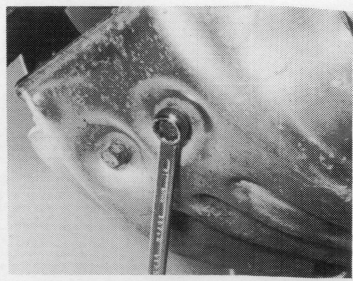
4.2b Where fitted, remove the bashplate

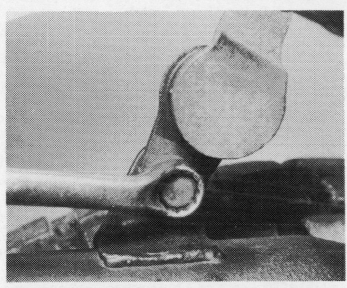
4.6a Exhaust removal TY175 – remove rear mounting bolt and withdraw tailpipe ...

4.6b ... then unscrew front mounting nuts ...

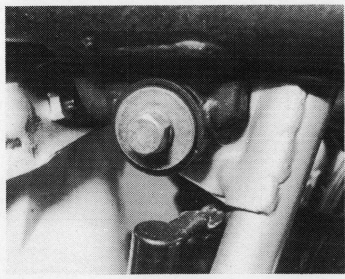
4.6c ... and main mounting bolt

4.7 Mark shaft before removing kickstart lever

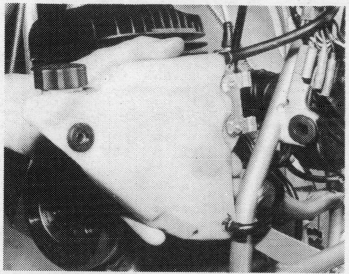

4.8 Disconnect oil feed pipe – tank can be removed for extra working space

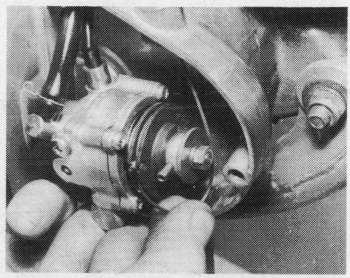

4.9 Disconnect oil pump cable from pump pulley

4.11a Flatten nipple retaining tang to release clutch cable from operating lever

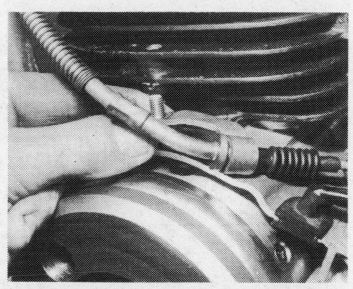

4.11b Clutch cable bracket can be unbolted from barrel – TY125 and 175

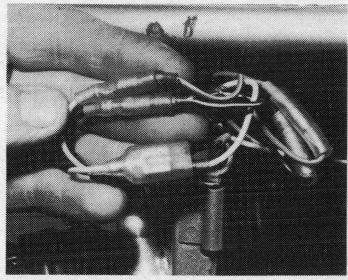

4.13 Disconnect all electrical wiring, noting colour coding

4.14a Remove engine mounting bolts ...

4.14b ... noting that front mountings (TY125 and 175) have special washers

5 Dismantling the engine/gearbox unit: preliminaries

1 Before any dismantling work is undertaken, the external surfaces of the unit should be thoroughly cleaned and degreased. This will prevent the contamination of the engine internals, and will also make working a lot easier and cleaner. A high flash point solvent, such as paraffin (kerosene) can be used, or better still, a proprietary engine degreaser such as Gunk. Use old paintbrushes and toothbrushes to work the solvent into the various recesses of the engine castings. Take care to exclude solvent or water from the electrical components and inlet and exhaust ports. The use of petrol (gasoline) as a cleaning medium should be avoided, because the vapour is explosive and can be toxic if used in a confined space.
2 When clean and dry, arrange the unit on the workbench, leaving a suitable clear area for working. Gather a selection of small containers and plastic bags so that parts can be grouped together in an easily identifiable manner. Some paper and a pen should be on hand to permit notes to be made and labels attached where necessary. A supply of clean rag is also required.
3 Before commencing work, read through the apropriate section so that some idea of the necessary procedure can be gained. When removing the various engine components it should be noted that great force is seldom required, unless specified. In many cases, a component's reluctance to be removed is indicative of an incorrect approach or removal method. If in any doubt, re-check with the text.

6 Dismantling the engine/gearbox unit: removing the cylinder head, barrel and piston

1 These components can be removed with the engine/gearbox unit in or out of the frame, but in the former case the HT lead must be disconnected from the spark plug, and the carburettor and exhaust pipe must be removed. Also, on TY125/175 models, the clutch cable bracket must be unbolted from the cylinder barrel; on TY50/80 models, disconnect the oil feed pipe.
2 Unscrew the spark plug and then remove the four bolts securing the intake stub/reed valve assembly to the rear of the barrel. Tapping very gently with a soft-faced mallet to break the seal of the two gaskets, lift away the intake stub and the reed valve casing.
3 Working in a diagonal sequence and by one turn at a time to avoid the risk of distortion, remove the four cylinder head retaining nuts (and lock washers, where fitted), then lift away the head and its gasket.
4 Bring the piston to the top of its stroke, then lift the barrel just enough to expose the bottom of the piston skirt. It may be necessary to use a soft-faced mallet to tap gently around the barrel to break the seal of the base gasket; take great care not to damage the fins. Pack

a wad of clean rag in the crankcase mouth to prevent dirt or debris from dropping in, then lift the barrel off the piston.
5 Use a sharp-pointed instrument or a pair of needle-nose pliers to remove one of the gudgeon pin retaining circlips, press out the gudgeon pins far enough to clear the connecting rod and withdraw the piston. Push out the small-end bearing. Discard the used circlips and obtain new ones for reassembly.
6 If the gudgeon pin is a tight fit in the piston, soak a rag in boiling water, wring it out and wrap it around the piston; the heat will expand the piston sufficiently to release its grip on the pin. If necessary, the pin may be tapped out using a hammer and drift, but take care to support firmly the piston and connecting rod while this is done.
7 The piston rings are removed by holding the piston in both hands and prising the ring ends apart gently with the thumbnails until the rings can be lifted out of their grooves and on to the piston lands, one side at a time. The rings can then be slipped off the piston and put to one side for cleaning and examination. If the rings are stuck in their grooves by excessive carbon deposits use three strips of thin metal sheet to remove them, as shown in the accompanying illustration. Be careful, as the rings are brittle and will break easily if overstressed.

6.4 Pack crankcase mouth with clean rag before fully removing barrel

7 Dismantling the engine/gearbox unit: removing the crankcase right-hand cover

1 This may be done with the engine/gearbox unit in or out of the frame but in the former case the transmission oil must be drained (or the machine laid down on its left-hand side so that the oil does not spill out), the oil pump cable and feed pipes must be disconnected, the kickstart lever must be removed and the rear brake adjustment slackened off so that the pedal can be depressed clear of the cover. On TY125 and TY175 models the crankcase bashplate must be removed, and on TY80 models the swinging arm pivot bolt retaining nut must be removed and the footrest mounting bolt slackened so that the footrest can be pivoted backwards clear of the cover. Refer to Section 4 of this Chapter.
2 Note that the oil pump and drive can remain in place on the cover; if their removal is required refer to Chapter 2.
3 Working progressively and in a diagonal sequence from the outside in, slacken the cover retaining screws and withdraw them. Store them in a cardboard template of the cover to provide a guide to their correct positions on reassembly.
4 Pull the cover away, tapping gently with a soft-faced mallet to break the seal. Be careful that there are no thrust washers adhering to the cover. Peel off the gasket; unless the two locating dowels are firmly fixed in the crankcase, these should be removed and stored with the cover.

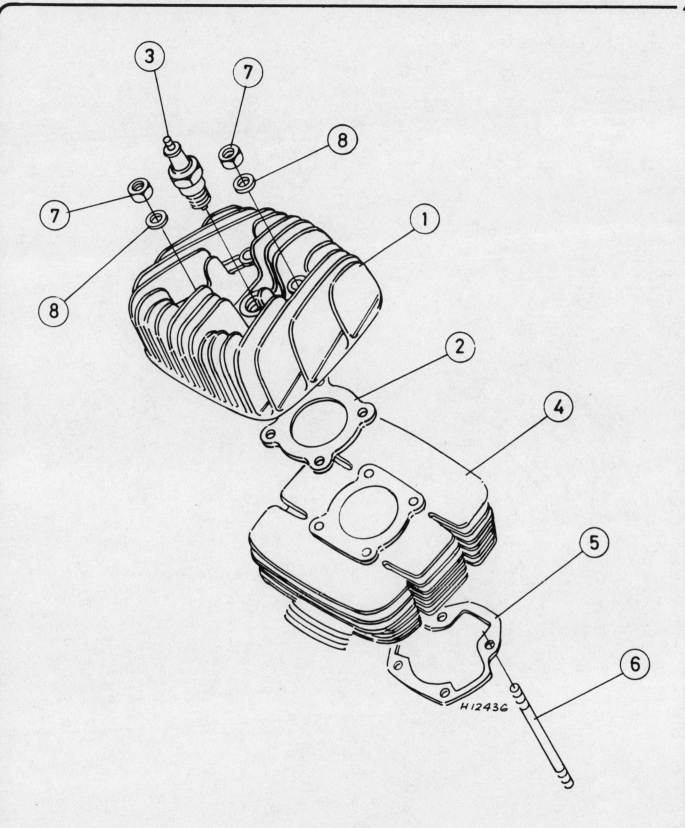

H12436

Fig. 1.1 Cylinder head and barrel – TY50

1	Cylinder head	5	Base gasket
2	Cylinder head gasket	6	Stud
3	Spark plug	7	Nut
4	Cylinder barrel	8	Washer

Fig. 1.2 Cylinder head and barrel – TY80

1 Cylinder head
2 Cylinder head gasket
3 Spark plug
4 Nut – 4 off
5 Cylinder barrel
6 Base gasket
7 Stud

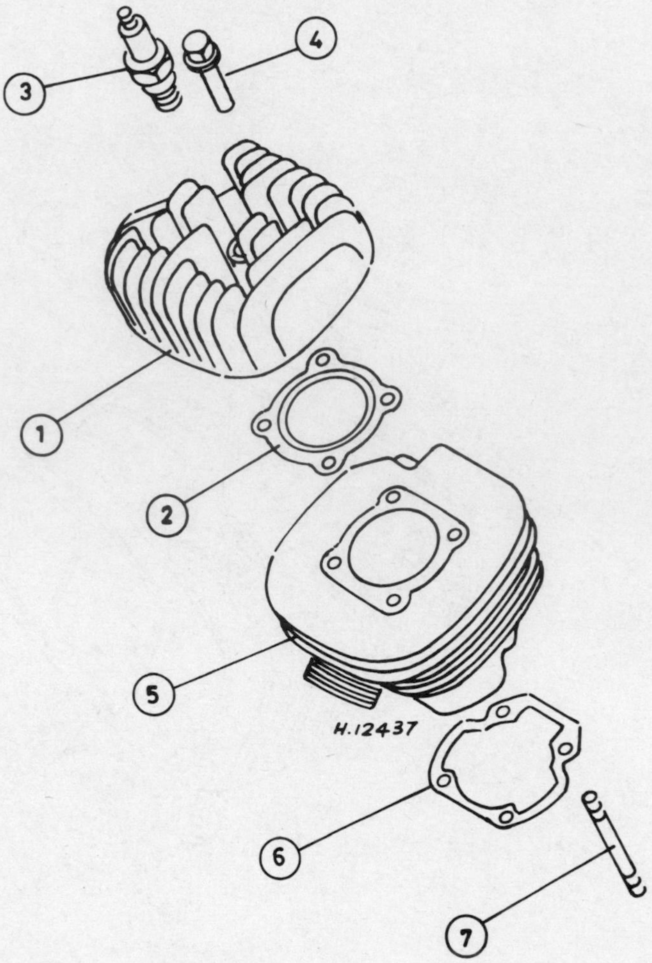

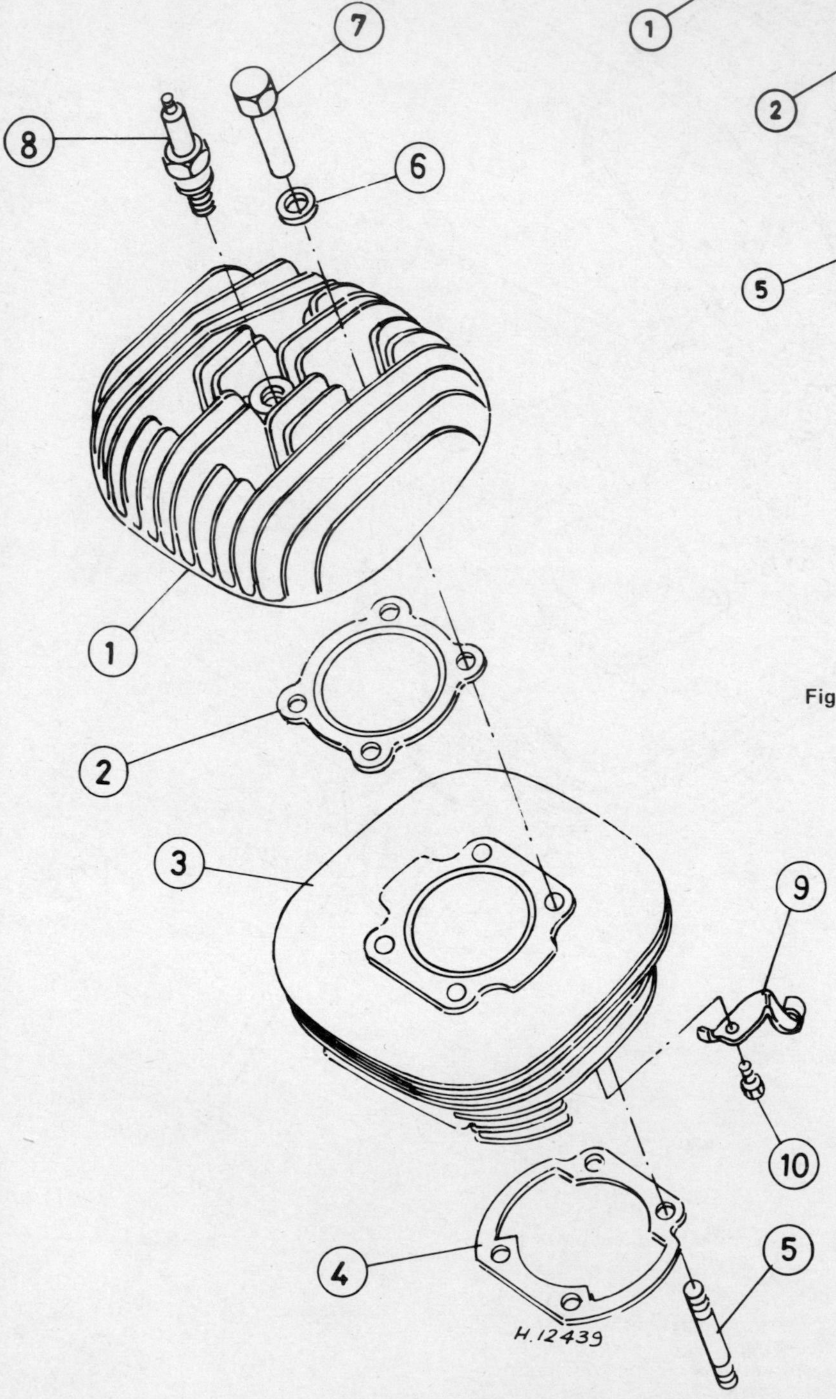

H.12437

H.12439

Fig. 1.3 Cylinder head and barrel – TY125 and 175

1 Cylinder head
2 Cylinder head gasket
3 Cylinder barrel
4 Base gasket
5 Stud
6 Washer
7 Nut – 4 off
8 Spark plug
9 Cable guide
10 Bolt

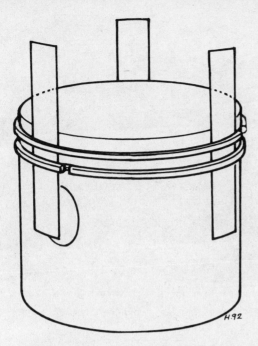

Fig. 1.4 Removing gummed piston rings

8 Dismantling the engine/gearbox unit: removing the clutch and primary drive pinion

1   These components can be removed after the crankcase right-hand cover has been withdrawn, as described in the previous Section.

2   Avoid distortion of the clutch pressure plate by slackening its retaining bolts evenly and in a diagonal sequence. Remove the bolts, washers and springs, the pressure plate, the friction and plain plates and the cushion rings (where fitted). Withdraw the two lengths of pushrod and the ball bearing from the gearbox input shaft, using a magnet to extract them.

3   The clutch centre must be locked to permit the retaining nut to be slackened. First flatten back, where applicable, the raised portion of the retaining nut lock washer. If the engine is in the frame, select top gear and apply the rear brake hard to lock the transmission. Alternatively, make up a clutch holding tool from two strips of metal and a nut and bolt; as shown in the accompanying photograph. With the clutch centre locked, remove the retaining nut and its lock washer.

4   Lift away the clutch centre, noting the exact number and position of any thrust washers behind it, then slide off the outer drum, followed by the bearing sleeve and large thrust washer.

5   Lock the crankshaft by passing a close-fitting bar through the small-end eye. Place wooden blocks between the crankcase mouth and bar. Alternatively sprag the primary drive gears by wedging between them a piece of soft metal, wood or a tightly-wadded rag. Remove the drive pinion retaining nut, its lock washer, the pinion and Woodruff key and the spacer.

6   On TY125 and TY175 models, slacken the locknut and remove fully the adjuster screw, then pull the clutch release shaft with its return spring and plain washer out of the crankcase left-hand half.

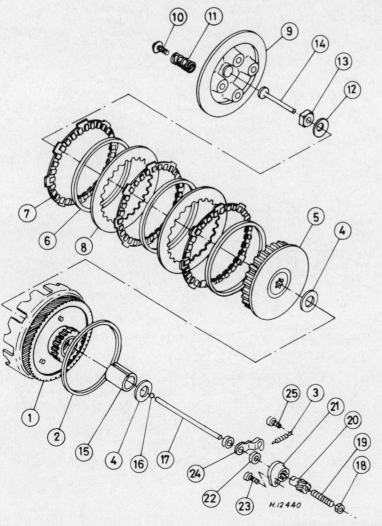

Fig. 1.5 Clutch – TY50 and 80

1   *Outer drum*
2   *O-ring*
3   *Return spring*
4   *Thrust washer – 3 off early models, 2 off later models*
5   *Clutch centre*
6   *Cushion ring – 3 off early TY50, all TY80, 2 off late TY50*
7   *Friction plate – 3 off early TY50, all TY80, 2 off late TY50*
8   *Plain plate – 2 off early TY50, all TY80, 1 off late TY50*
9   *Pressure plate*
10 *Bolt – 4 off*
11 *Spring – 4 off*
12 *Lock washer*
13 *Nut*
14 *Headed push rod*
15 *Bearing sleeve*
16 *Steel ball*
17 *Plain push rod*
18 *Locknut*
19 *Adjusting screw*
20 *Release worm*
21 *Body*
22 *Oil seal*
23 *Screw*
24 *Release lever*
25 *Pin*

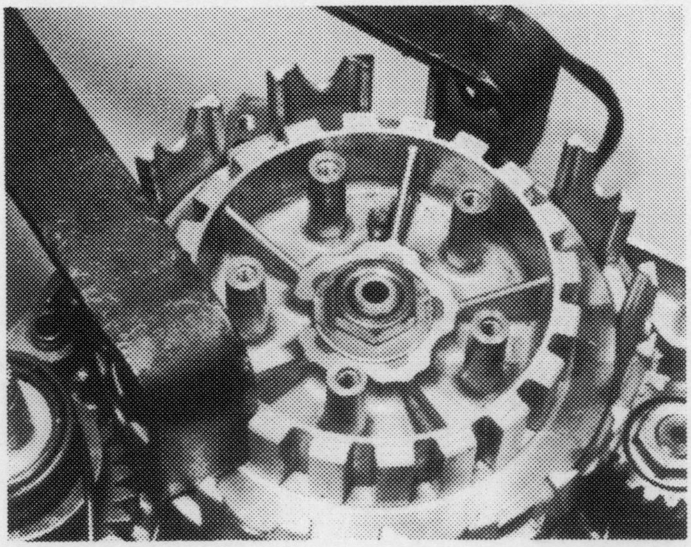

8.3 Special tool can be made up as shown to hold clutch centre while nut is removed

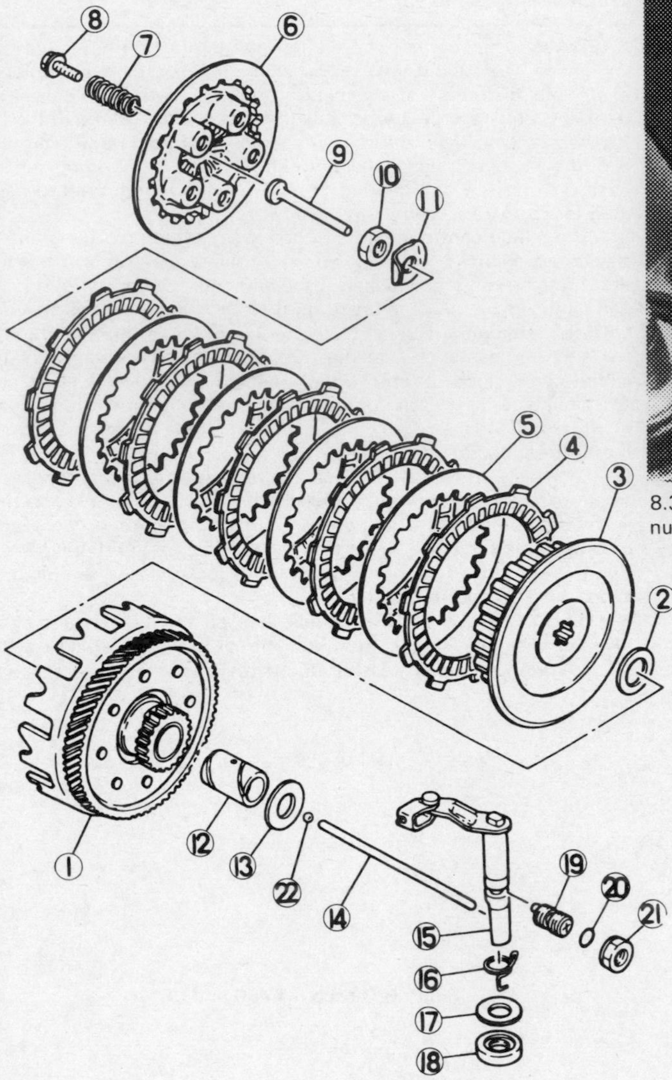
1
2
3
4
5
6
7
8
9
10
11
12
13
14
15
16
17
18
19
20
21
22

Fig. 1.6 Clutch – TY125 and 175

1	*Outer drum*	12	*Spacer*
2	*Thrust washer*	13	*Thrust washer*
3	*Clutch centre*	14	*Pushrod*
4	*Friction plate*	15	*Release lever*
5	*Plain plate*	16	*Return spring*
6	*Pressure plate*	17	*Washer*
7	*Spring*	18	*Oil seal*
8	*Bolt*	19	*Adjuster screw*
9	*Pushrod*	20	*O-ring*
10	*Nut*	21	*Nut*
11	*Lock washer*	22	*Steel ball*

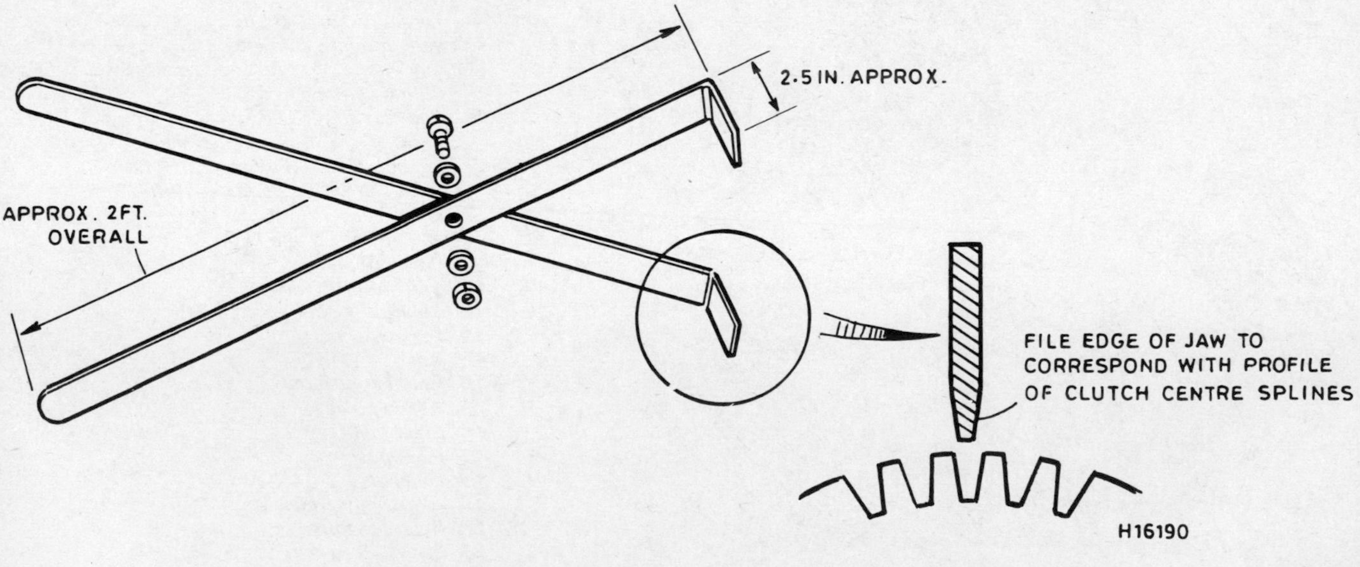
2·5 IN. APPROX.
APPROX. 2FT. OVERALL
FILE EDGE OF JAW TO CORRESPOND WITH PROFILE OF CLUTCH CENTRE SPLINES
H16190

Fig. 1.7 Fabricated clutch holding tool

9 Dismantling the engine/gearbox unit: removing the kickstart assembly

1 The kickstart can be removed only after the crankcase right-hand cover has been withdrawn as described in Section 7, but while the shaft assembly can be removed from behind the clutch, the idler gear cannot be withdrawn until the clutch has been removed as described in the previous Section.

2 Using a pair of heavy pliers, unhook the kickstart return spring from its stop and allow it to unwind slowly. Pull the complete shaft assembly clear of the crankcase. Refer to the accompanying line drawings when dismantling the shaft assembly.

3 Remove from the output shaft left-hand end the circlip, the plain washer, the idler pinion (noting which way round it is fitted), the second plain washer or wave washer and finally the plain washer or circlip (as appropriate to the model being worked on).

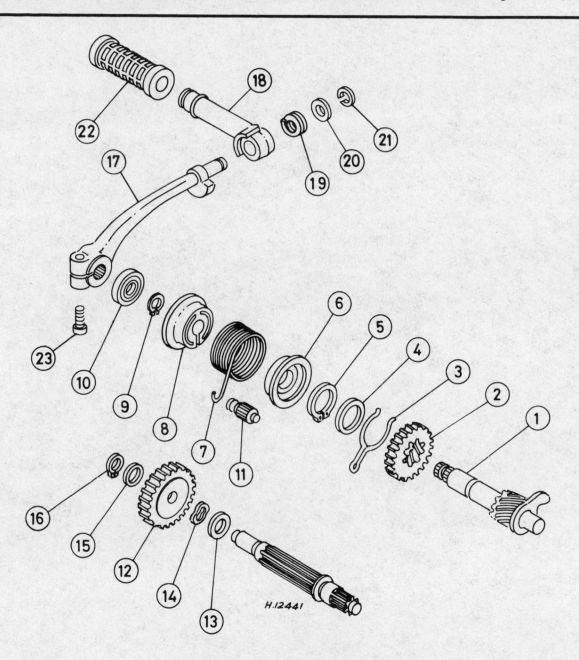

H.12441

Fig. 1.8 Kickstart assembly – TY50 and 80

1	Kickstart shaft	7	Return spring	13	Washer	19 Spring washer
2	Kickstart pinion	8	Spring guide	14	Wave washer	20 Washer
3	Friction clip	9	Circlip	15	Washer	21 Circlip
4	Thrust washer	10	Oil seal	16	Circlip	22 Lever rubber
5	Circlip	11	Return spring stop	17	Kickstart lever	23 Pinch bolt
6	Spring guide	12	Idler pinion	18	Kickstart lever knuckle	

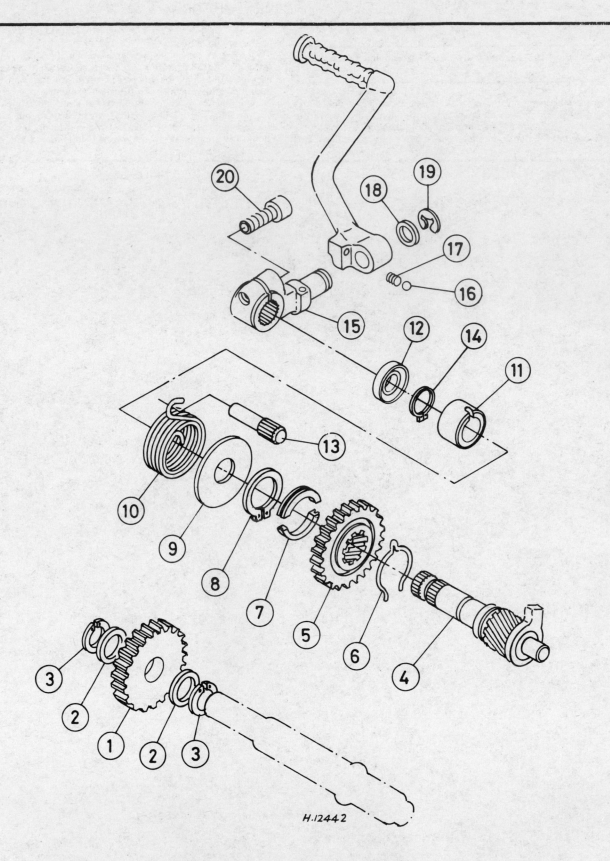

H.12442

Fig. 1.9 Kickstart assembly – TY125 and 175

1	Idler pinion	6	Friction clip	11	Spring guide	15 Kickstart lever knuckle
2	Washer – 2 off	7	Split collets	12	Oil seal	16 Ball
3	Circlip – 2 off	8	Circlip	13	Return spring stop	17 Spring
4	Kickstart shaft	9	Washer	14	Circlip – early 175	18 Washer
5	Kickstart pinion	10	Return spring		model	19 Circlip
						20 Allen bolt

10 Dismantling the engine/gearbox unit: removing the external gearchange components

General

1   These components can be removed with the engine/gearbox unit in or out of the frame, but in the former case the gearchange pedal must be removed and the crankcase right-hand cover and clutch must be withdrawn, as described in the earlier Sections of this Chapter.

2   Before removing the mechanism, press the gearchange pedal up or down and release it; in both directions the mechanism should return smoothly, quickly and without free play to the rest position. If not, or if any free play is discernible the gearchange shaft return spring is fatigued and must be renewed on reassembly. Similarly check the claw arm spring.

TY50, TY80

3   Having removed the gearchange pedal and left-hand crankcase cover, prise the circlip off the gearchange shaft left-hand end, then withdraw the plain washer.

4   Prise off the circlip retaining the selector claw arm assembly to its pivot, press the claw arm downwards clear of the selector drum and lift away the assembly. Lift the roller off the end of the gearchange shaft and pull the shaft out of the crankcase. Remove from the crankcase left-hand side the neutral detent plunger assembly.

TY125, TY175

5   Having removed the gearchange pedal and left-hand crankcase cover, remove the four screws which retain the linkage cover to the crankcase left-hand side, below the output shaft end. Remove the cover, noting its sealing O-ring and two locating dowel pins.

6   Prise off the circlip holding the linkage arm to the gearchange shaft, withdraw the thrust washer and pull the arm off the shaft splines, followed by the second thrust washer. Pull the roller off the gearchange pedal shaft and withdraw the pedal shaft, noting the thrust washers one on each side.

7   Working on the engine right-hand side, press upwards the claw arm clear of the selector drum and pull the gearchange shaft and return spring out of the crankcase. Disengage the detent roller arm spring from the bearing retaining plate, unscrew the pivot bolt and withdraw the roller arm assembly. Remove from the crankcase left-hand side the neutral detent plunger assembly.

All models

8   If required, use an impact driver to release its retaining screw and lift away the selector drum end cover, then lift out the selector pins with a pair of pliers.

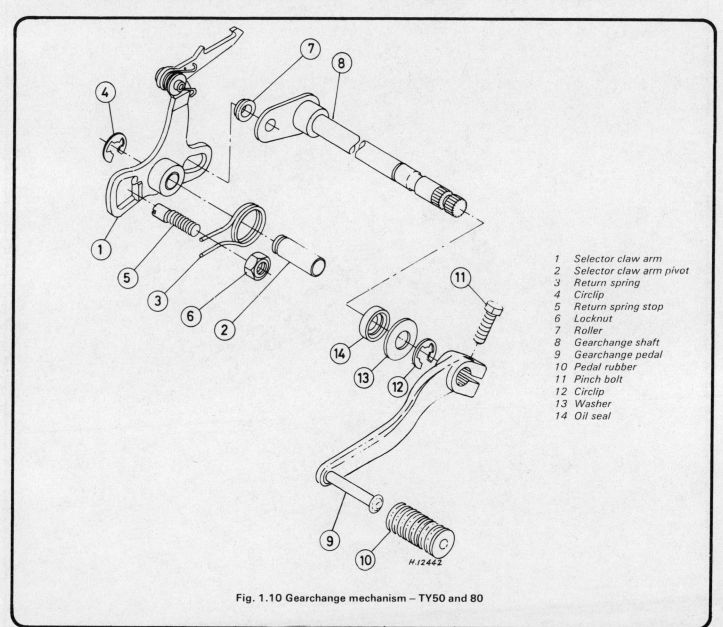

1   Selector claw arm
2   Selector claw arm pivot
3   Return spring
4   Circlip
5   Return spring stop
6   Locknut
7   Roller
8   Gearchange shaft
9   Gearchange pedal
10   Pedal rubber
11   Pinch bolt
12   Circlip
13   Washer
14   Oil seal

H.12442

Fig. 1.10 Gearchange mechanism – TY50 and 80

10.7 Removing the gearchange shaft – TY125 and 175

**Fig. 1.11 Gearchange mechanism –
TY125 and 175**

1 Gearchange pedal
2 Pinch bolt
3 Oil seal
4 Gearchange pedal shaft
5 Linkage rear arm
6 Thrust washer – 2 off
7 Roller
8 Linkage arm
9 Circlip
10 Thrust washer
11 Oil seal
12 Gearchange shaft
13 Spring
14 Return spring
15 Return spring stop
16 Locknut
17 Lock washer

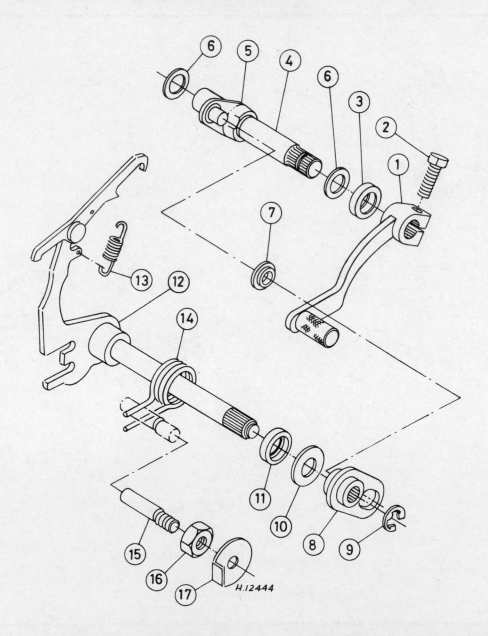

H.12444

11 Dismantling the engine/gearbox unit: removing the flywheel generator

1   Before the generator can be removed, the gearchange pedal and the crankcase left-hand cover must be withdrawn, as described in Section 4.

2   To lock the crankshaft to permit the removal of the rotor retaining nut, select top gear and apply the rear brake hard or apply a holding tool such as a strap wrench to the rotor itself. A more positive holding tool can be fabricated from two strips of metal, as shown in the accompanying illustration. If the top end has been dismantled, pass a close-fitting metal bar through the connecting rod small-end eye and support it on two wooden blocks placed across the crankcase mouth. With the crankshaft locked, remove the retaining nut and the washer(s) behind it.

3   Remove the rotor using only a centre bolt flywheel puller (see accompanying photograph). These are available from Yamaha Service Agents under various part numbers, but pattern versions to suit the majority of small-capacity Japanese machines are available at a lower price from most good motorcycle dealers. One of these tools should be considered essential; **no other method of rotor removal is recommended.**

4   Unscrew the tool centre bolt and screw the tool body anti-clockwise into the thread in the centre of the rotor (a **left-hand thread** is employed) until it seats firmly. Tighten the centre bolt down on to the crankshaft end and tap smartly on the bolt head with a hammer. The shock should jar the rotor free; if not tighten the centre bolt further and tap again. Withdraw the rotor and its Woodruff key from the crankshaft.

5   Where fitted, disconnect the lead from the neutral indicator switch. Unclip the generator lead from the crankcase, note the stator fitted position and remove its two countersunk screws; lift away the stator plate.

6   If required, on machines so equipped, the neutral indicator switch can be unscrewed from the crankcase.

11.2 Note method used to lock crankshaft while rotor retaining nut is unscrewed

11.3 Pattern flywheel puller in use – no other method of rotor removal is recommended

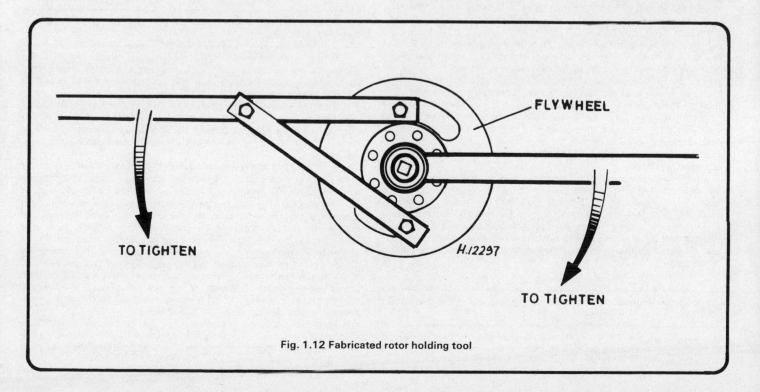

FLYWHEEL

TO TIGHTEN

TO TIGHTEN

H.12297

Fig. 1.12 Fabricated rotor holding tool

11.5 Stator plate is located by mounting screws – no need to make timing marks on removal

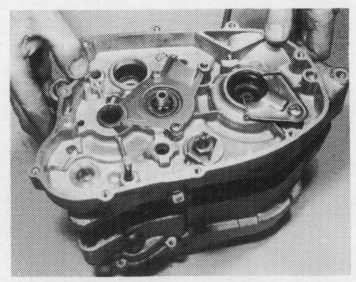

12.3 TY125 and 175 models – lift right-hand crankcase half away – reverse applies to 50 and 80 models

12 Dismantling the engine/gearbox unit: separating the crankcase halves

1   The crankcases can be separated only after the engine/gearbox unit has been removed from the frame and all preliminary dismantling operations described in Sections 4-11 of this Chapter have been carried out. On TY50 and TY80 models, use an impact driver to release the selector drum guide plate securing screws, then remove the plate; remove also the gearbox sprocket spacer.
2   Make a final check that all components have been removed which might hinder crankcase separation. Make a cardboard template marking their respective positions and remove all crankcase fastening screws, slackening them progressively and in a diagonal sequence from the outside inwards.
3   On TY50 and TY80 models the crankcase left-hand half is to be lifted away leaving the crankshaft and gearbox components in the right-hand half. On TY125 and TY175 models the reverse is the case, ie the right-hand half is lifted away, leaving all components in the left-hand half.
4   Using only a soft-faced mallet, tap gently on the exposed ends of the crankshaft and gearbox shafts and all around the joint area of the two crankcase halves until initial separation is achieved. Lift off the half to be removed, ensuring it remains absolutely square so that the bearings do not stick on their respective shafts; tap gently on the shaft ends to assist removal. **Do not** use excessive force and **never** attempt to lever the cases apart. If undue difficulty is encountered, tap the cases back together and start again. If all else fails, take the assembly to a Yamaha Service Agent for the cases to be separated using special tool Part Number 90890-01135. This tool can also be used to press out the crankshaft.
5   If difficulty is encountered in achieving initial separation, it may be because corrosion has formed on the locating dowel pins. Apply a quantity of penetrating fluid to the joint area and inside the various mounting bolt passages, allow time for it to work and start again.
6   When the casing is removed, check that there are no loose components such as dowels or thrust washers which might drop clear and be lost. Any such components should be refitted in their correct locations.

13 Dismantling the engine/gearbox unit: removing the crankshaft and gearbox components

1   On TY125 and TY175 models, lift out the shaft holding the single selector fork in front of the selector drum, then remove the gearbox sprocket spacer.
2   On all models, hold both gearbox clusters, the selector drum and

forks as a single assembly in one hand and pull them out of the crankcase while tapping gently with the soft-faced mallet on the exposed ends of the shafts (and selector drum, if applicable). Be very careful not to lose any small thrust washers or similar components which might drop clear.
3   Firmly support the crankcase half with the crankshaft on wooden blocks so that the crankshaft end is clear of the work surface. Protect the threaded end of the shaft by refitting the appropriate nut, place a copper drift against the shaft and strike it with a heavy hammer to drift the shaft down and clear of the crankcase. Do not use excessive force, a few firm blows should suffice. Be careful not to allow the crankshaft to drop clear; a thick layer of rag should be placed on the work surface underneath it.
4   If all else fails do not resort to excessive force, but take the assembly to a Yamaha Service Agent for the crankshaft to be pressed out using service tool Part Number 90890-01135.

14 Dismantling the engine/gearbox unit: removing oil seals and bearings

1   Before removing any oil seal or bearing, check that it is not secured by a retaining plate. If this is the case, use an impact driver or spanner, as appropriate, to release the securing screws or bolts and lift away the retaining plate.
2   Oil seals are easily damaged when disturbed and, thus, should be renewed as a matter of course during overhaul. Prise them out of position using the flat of a screwdriver and taking care not to damage the alloy seal housings; take care to note which way round the seals are fitted.
3   It is possible to remove the main bearing oil seals without separating the crankcases; either by screwing in two self-tapping screws so that the seal can be pulled out with two pairs of pliers, or by simply digging the seal out with a sharply-pointed instrument. This method is not recommended as it is difficult to carry out without scratching the seal housing or crankshaft or without damaging one or both of the main bearings. Furthermore, it is almost impossible to fit a new seal without damaging it and nothing can be done to trace and rectify the fault which caused the seal to fail in the first place.
4   The crankshaft and gearbox bearings are a press fit in their respective crankcase locations. To remove a bearing, the crankcase casting must be heated so that it expands and releases its grip on the bearing, which can be drifted or pulled out.
5   To prevent casting distortion, it must be heated evenly to a temperature of about 100°C by placing it in an oven; if an oven is not available, place the casting in a suitable container and carefully pour boiling water over it until it is submerged.
6   Taking care to prevent personal injury when handling heated

components, lay the casting on a clean surface and tap out the bearing using a hammer and a suitable drift. If the bearing is to be re-used apply the drift only to the bearing outer race, where this is accessible, to avoid damaging the bearing. In some cases it will be necessary to apply pressure to the bearing inner race; in such cases closely inspect the bearing for signs of damage before using it again. When drifting a bearing from its housing it must be kept square to the housing to prevent tying in the housing with the resulting risk of damage. Where possible, use a tubular drift such as a socket spanner which bears only on the bearing outer race; if this is not possible, tap evenly around the outer race to achieve the same result.

7   In some cases, bearings are pressed into blind holes in the castings. These bearings must be removed by heating the casting and tapping it face downwards on to a clean wooden surface to dislodge the bearing under its own weight. If this is not successful the casting should be taken to a motorcycle service engineer who has the correct internally expanding bearing puller. If a bearing sticks to its shaft on removal, it can only be safely removed using a knife-edged bearing puller.

14.1 Removing oil seals – take care not to damage housing

15 Examination and renovation: general

1   Before examining the parts of the dismantled engine unit for wear it is essential that they should be cleaned thoroughly. Use a petrol/paraffin mix or a high flash-point solvent to remove all traces of old oil and sludge which may have accumulated within the engine. Where petrol is included in the cleaning agent normal fire precautions should be taken and cleaning should be carried out in a well ventilated place.

2   Examine the crankcase castings for cracks or other signs of damage. If a crack is discovered it will require a specialist repair.

3   Examine carefully each part to determine the extent of wear, checking with the tolerance figures listed in the Specifications section of this Chapter or in the main text. If there is any doubt about the condition of a particular component, play safe and renew.

4   Use a clean lint-free rag for cleaning and drying the various components. This will obviate the risk of small particles obstructing the internal oilways, and causing the lubrication system to fail.

5   Various instruments for measuring wear are required, including a vernier gauge or external micrometer and a set of standard feeler gauges. Both an internal and external micrometer will be required to check wear limits. Additionally, although not absolutely necessary, a dial gauge and mounting bracket are invaluable for accurate measurement of end float, and play between components of very low diameter bores – where a micrometer cannot reach. After some experience has

been gained, the state of wear of many components can be determined visually or by feel and thus a decision on their suitability for continued service can be made without resorting to direct measurement.

16 Examination and renovation: engine cases and covers

1   Small cracks or holes in aluminium castings may be repaired with an epoxy resin adhesive, such as Araldite, as a temporary measure. Permanent repairs can only be effected by argon-arc welding, and only a specialist in this process is in a position to advise on the economics or practicability of such a repair.

2   Damaged threads can be economically reclaimed by using a diamond section wire insert, of the Helicoil type, which is easily fitted after drilling and re-tapping the affected thread. Most motorcycle dealers and small engineering firms offer a service of this kind.

3   Sheared studs or screws can usually be removed with screw extractors, which consist of tapered, left-hand thread screws, of very hard steel. These are inserted by screwing anti-clockwise into a pre-drilled hole in the stud, and usually succeed in dislodging the most stubborn stud or screw. If a problem arises which seems to be beyond your scope, it is worth consulting a professional engineering firm before condemning an otherwise sound casting. Many of these firms advertise regularly in the motorcycle papers.

17 Examination and renovation: bearings and oil seals

1   The crankshaft and gearbox bearings can be examined while they are still in place in the crankcase castings. Wash them thoroughly to removal all traces of oil, then feel for free play by attempting to move the inner race up and down, then from side-to-side. Examine the bearing balls or rollers and the bearing tracks for pitting or other signs of wear, then spin the bearing hard. Any roughness caused by defects in the bearing balls or rollers or in the bearing tracks will be felt and heard immediately.

2   If any signs of free play are discovered, or if the bearing is not free and smooth in rotation but runs roughly and slows down jerkily, it must be renewed. Bearing removal is described in Section 14 of this Chapter, and refitting in Section 28.

3   To prevent oil leaks occurring in the future, all oil seals and O-rings should be renewed whenever they are disturbed during the course of an overhaul, regardless of their apparent condition. This is particularly true of the main bearing oil seals, which are a weak point on any two-stroke.

18 Examination and renovation: cylinder head

1   Check that the cylinder head fins are not clogged with oil or road dirt, otherwise the engine will overheat. If necessary, use a degreasing agent and brush to clean between the fins. Check that no cracks are evident, especially in the vicinity of the spark plug or stud holes.

2   Check the condition of the thread in the spark plug hole. If it is damaged an effective repair can be made using a Helicoil thread insert. This service is available from most motorcycle dealers. Always use the correct plug and do not overtighten, see Routine Maintenance.

3   Leakage between the head and barrel will indicate distortion. Check the head by placing a straight-edge across several places on its mating surface and attempting to insert a 0.02 mm (0.0008 in) feeler gauge between the two.

4   Remove excessive distortion by rubbing the head mating surface in a slow circular motion against emery paper placed on plate glass. Start with 200 grade paper and finish with 400 grade and oil. Do not remove an excessive amount of metal. If in doubt consult a Yamaha agent.

5   Note that most cases of cylinder head distortion can be traced to unequal tensioning of the cylinder head securing nuts or to tightening them in the incorrect sequence.

6   Plain copper cylinder head gaskets are the only type of gasket that can be reused. If such a gasket is undamaged, it should be annealed (softened) by heating to a cherry red and quenching it in cold water. Take care to prevent personal injury when doing this.

19 Examination and renovation: cylinder barrel

1   The usual indication of a badly worn cylinder barrel and piston is piston slap, a metallic rattle that occurs when there is little or no load on the engine.

2   Clean all dirt from between the cooling fins. Carefully remove the ring of carbon from the bore mouth so that bore wear can be accurately assessed and check the barrel/cylinder head mating surface as described in the previous Section. Clean all carbon from the exhaust port and all traces of old gasket from the cylinder base.

3   Examine the bore for scoring or other damage, particularly if broken rings are found. Damage will necessitate reboring and a new piston regardless of the amount of wear. A satisfactory seal cannot be obtained if the bore is not perfectly finished.

4   There will probably be a lip at the uppermost end of the cylinder bore which marks the limit of travel of the top of the piston ring. The depth of the lip will give some indication of the amount of bore wear that has taken place even though the amount of wear is not evenly distributed.

5   The most accurate method of measuring bore wear is by the use of a cylinder bore DTI (Dial Test Indicator) or a bore micrometer. Measure at the top (just below the wear ridge), middle and bottom of the bore, in line with the gudgeon pin axis and at 90° to it. Take six measurements in all noting each one carefully. None of the measurements in any plane should differ from its counterparts by more than 0.05 mm (0.002 in) ie the maximum taper tolerance, and none of those measurements at 90° to the gudgeon pin should exceed their fellows along its axis by more than the maximum ovality tolerance specified. If either tolerance is exceeded at any point, the barrel must be rebored and an oversize piston and rings fitted. This can be confirmed by subtracting the overall diameter of the piston from the minimum cylinder bore figure obtained. If the difference calculated exceeds 0.1 mm (0.004 in) the bore can be confirmed as excessively worn, assuming that the piston is unworn (See Section 20).

6   Alternatively, insert the piston into the bore in its correct position with the skirt just below the wear ridge, ie at the point of maximum wear. Measure the gap using feeler gauges. If the piston/cylinder clearance is 0.1 mm (0.004 in) or more the barrel must be rebored. It must be stressed that this can only be a guide to the degree of bore wear alone unless the piston is known to be unworn; have the barrel measured accurately by a competent Yamaha Service Agent or similar expert before a rebore is undertaken.

7   After a rebore has been carried out, the reborer should hone the bore lightly to provide a fine cross-hatched surface so that the new piston and rings can bed in correctly. Also the edges of the ports should be chamfered first with a scraper, then with fine emery, to prevent the rings from catching on them and breaking.

8   If a new piston and/or rings are to be run in a part-worn bore the surface should be prepared first by glaze-busting. This involves the use of a special honing attachment with (usually) an electric drill to provide a surface similar to that described above. Most motorcycle dealers have such equipment and will be able to carry out the necessary work for a small charge.

20 Examination and renovation: piston and piston rings

1   Disregard the existing piston and rings if a rebore is necessary; they will be replaced with oversize items. Note also that it is considered a worthwhile expense by many mechanics to renew the piston rings as a matter of course, regardless of their apparent condition.

2   Measure the piston diameter at right angles to the gudgeon pin axis, at a point 10 mm (0.4 in) above the base of the skirt. Since the manufacturer does not give service limits, the degree of piston wear can be assessed only by direct comparison with a new component. Ensure that the new component is the same size (standard or oversize) as the original and note that standard pistons are available in two sizes, the actual diameter being stamped in the piston crown.

3   Note that the manufacturer's alternative method of determining whether a rebore is necessary is to subtract the piston overall diameter from the minimum bore internal diameter measurement obtained. Since it is possible for the piston to have worn at a greater rate than

the bore, this can only be accurate if the piston measurement is that of an unworn component (ie new). If this method is used, always have your findings checked by an expert before undertaking a rebore; it may suffice to renew the piston alone.

4   Piston wear usually occurs at the skirt, especially on the forward face, and takes the form of vertical score marks. Reject any piston which is badly scored or which has been badly blackened as the result of the blow-by of gas. Slight scoring of the piston can be removed by careful use of a fine swiss file. Use chalk to prevent clogging of the file teeth and the subsequent risk of scoring. If the ring locating pegs are loose or worn, renew the piston.

5   The gudgeon pin should be a firm press fit in the piston. Check for scoring on the bearing surfaces of each part and where damage or wear is found, renew the part affected. The pin circlip retaining grooves must be undamaged; renew the piston rather than risk damage to the bore through a circlip becoming detached. Discard the circlips themselves; these should **never** be reused.

6   Any build-up of carbon in the ring grooves can be removed by using a section of broken piston ring, the end of which has been ground to a chisel edge. Using a feeler gauge, measure each ring to groove clearance, where applicable. Renew the piston if the measurement obtained exceeds that given in Specifications and if the rings themselves appear unworn.

7   Measure ring wear by inserting each ring into part of the bore which is unworn and measuring the gap between the ring ends with a feeler gauge. If the measurement exceeds that given in Specifications, renew the ring. Use the piston crown to locate the ring squarely in the bore approximately 20 mm (0.8 in) from the gasket surface.

8   Reject any rings which show discoloured patches on their mating surfaces; these should be brightly polished from firm contact with the cylinder bore.

9   Do not assume when fitting new rings that their end gaps will be correct; the installed end gap must be measured as described above to ensure that it is within the specified tolerances; if the gap is too wide another piston ring set must be obtained (having checked again that the bore is within specified wear limits), but if the gap is too narrow it must be widened by the careful use of a fine file.

21 Examination and renovation: crankshaft assembly

1   Big-end failure is characterised by a pronounced knock which will be most noticeable when the engine is worked hard. The usual causes of failure are normal wear, or failure of the lubrication supply. In the case of the latter, the noise will become apparent very suddenly, and will rapidly worsen.

2   Check for wear with the crankshaft set in the TDC (top dead centre) position, by pushing and pulling the connecting rod. No discernible movement will be evident in an unworn bearing, but care must be taken not to confuse end float, which is normal, and bearing wear.

3   If a dial gauge is available, set its pointer against the end of the small-end eye. Measure the side-to-side deflection of the connecting rod. Renew the big-end bearing if the measurement exceeds 2.0 mm (0.08 in).

4   Push the connecting rod to one side and use a feeler gauge to measure the big-end clearance. Renew the big-end bearing, the crank pin and its washers and the connecting rod if the clearance exceeds that specified.

5   Set the crankshaft on V-blocks positioned on a completely flat surface and measure the amount of deflection with the dial gauge pointer set against the inboard end of either mainshaft. Renew the crankshaft if the measurement exceeds 0.03 mm (0.0012 in).

6   Push the small-end bearing into the connecting rod eye and push the gudgeon pin through the bearing. Hold the rod steady and feel for movement between it and the pin. If movement is felt, renew the pin, bearing or connecting rod as necessary so no movement exists. Renew the bearing if its roller cage is cracked or worn.

7   Do not attempt to dismantle the crankshaft assembly, this is a specialist task. If a fault is found, return the assembly to a Yamaha agent who will supply a new or service-exchange item.

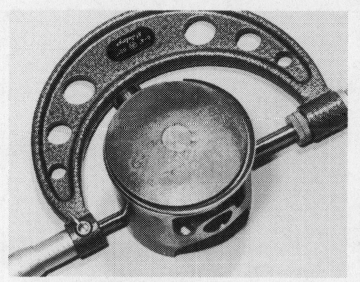

20.2 Measuring piston diameter

20.7 Measuring piston ring installed end gap

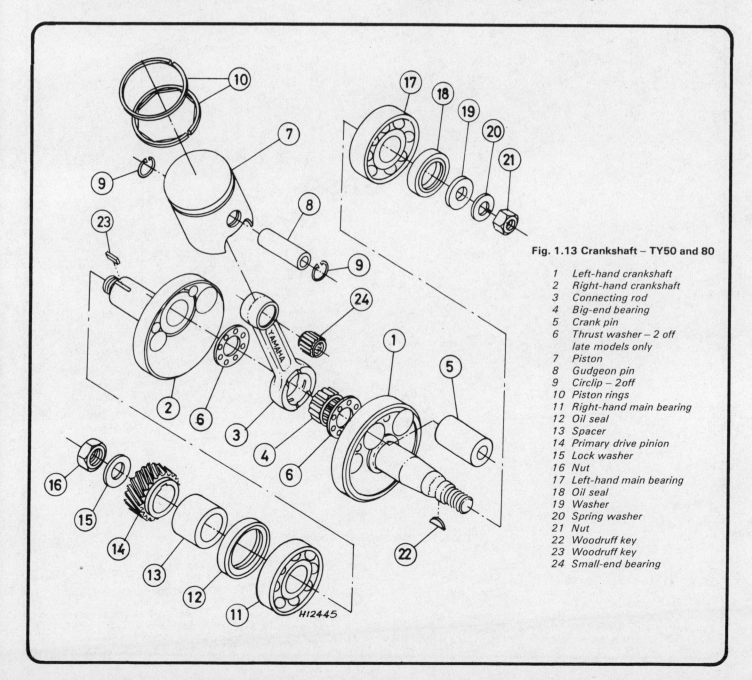

Fig. 1.13 Crankshaft – TY50 and 80

1   Left-hand crankshaft
2   Right-hand crankshaft
3   Connecting rod
4   Big-end bearing
5   Crank pin
6   Thrust washer – 2 off
    late models only
7   Piston
8   Gudgeon pin
9   Circlip – 2 off
10 Piston rings
11 Right-hand main bearing
12 Oil seal
13 Spacer
14 Primary drive pinion
15 Lock washer
16 Nut
17 Left-hand main bearing
18 Oil seal
19 Washer
20 Spring washer
21 Nut
22 Woodruff key
23 Woodruff key
24 Small-end bearing

H12445

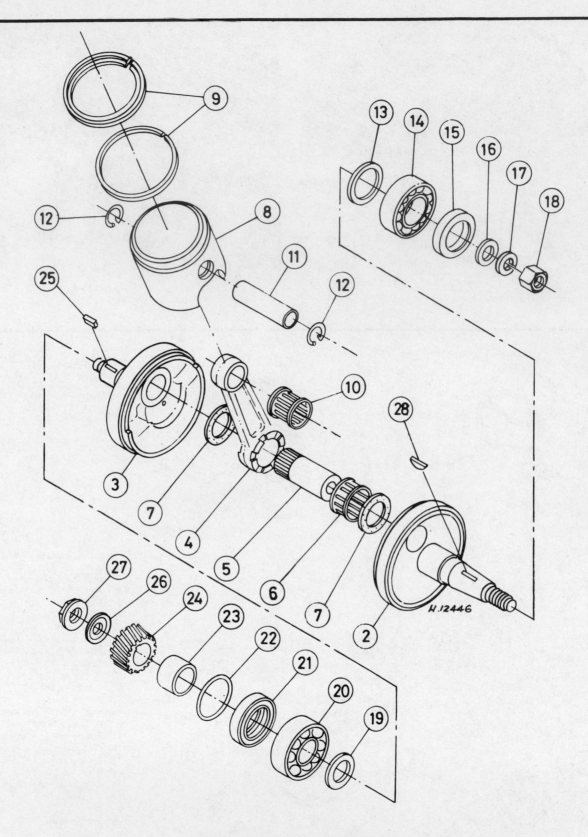

Fig. 1.14 Crankshaft – TY125 and 175

1	Crankshaft assembly	8	Piston	15	Oil seal	22	O-ring
2	Left-hand crankshaft	9	Piston rings	16	Washer	23	Spacer
3	Right-hand crankshaft	10	Small-end bearing	17	Spring washer	24	Primary drive pinion
4	Connecting rod	11	Gudgeon pin	18	Nut	25	Woodruff key
5	Crank pin	12	Circlip – 2 off	19	Shim – as required	26	Lock washer
6	Big-end bearing	13	Shim – as required	20	Right-hand main bearing	27	Nut
7	Thrust washer – 2 off	14	Left-hand main bearing	21	Oil seal	28	Woodruff key

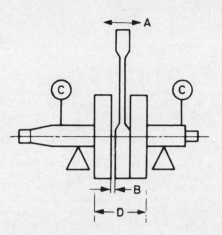

Fig. 1.15 Crankshaft measurement points

A   Big-end radial clearance   D   Crankshaft width
B   Big-end axial clearance   across flywheels
C   Crankshaft runout

22 Examination and renovation: primary drive

1   The primary drive consists of a crankshaft pinion which engages a large gear mounted on the inner face of the clutch drum. Both components are relatively lightly loaded and will not normally wear until a very high mileage has been covered.

2   If wear or damage is discovered it will be necessary to renew the component concerned. In the case of the large driven gear it will be necessary to purchase a complete clutch drum because the two items form an integral unit and cannot be obtained separately. Note that the large driven gear/clutch outer drum assembly has a shock absorber fitted. Check for play in this by holding the clutch outer drum and attempting to twist or rotate the primary driven gear backwards and forwards. Unfortunately no figures are given with which to assess the state of the shock absorber unit, and it will therefore be a matter of experience to decide whether renewal is necessary or not. Seek an expert opinion if any doubt exists about the amount of play discovered.

3   When obtaining new primary drive parts note that the two components are matched to give a prescribed amount of backlash. To this end, ensure that the match marks etched on the inner face of each are similar to avoid excessive or insufficient clearance. The marks, in the form of the letters 'B', 'C', or 'D' on TY50 and TY80 models, or 'A', 'B', 'C', 'D' or 'E' on TY125 and TY175 models must be the same on both components.

23 Examination and renovation: clutch

1   Clean the clutch components in a petrol/paraffin mix whilst observing the necessary fire precautions. Renew the O-ring fitted to the drum (TY50 and TY80 models only).

2   Overworn friction plates will cause clutch slip. Measure the thickness of each plate and renew it if the thickness is less than the limit given in Specifications. On TY50 and TY80 models, renew the cushion rings if any are damaged or worn.

3   Check the friction plate tangs and the drum slots for indentations caused by clutch chatter. If slight, the damage can be removed with a fine file, otherwise renewal is necessary.

4   Check all plates for distortion. Lay each one on a sheet of plate glass and attempt to insert a feeler gauge beneath it. Refer to Specifications for maximum allowable warpage.

5   Inspect each plain plate for scoring and signs of overheating in the form of blueing. Check the plate thickness with that given in Specifications. Remove slight damage to the plate tangs and hub slots with a fine file, otherwise renewal is necessary.

6   Examine the pressure plate for cracks and overheating. Look for hairline cracks around the base of each spring location and around the centre boss. Check for excessive distortion.

7   Measure the free length of each spring and renew all of them if one has set to less than the limit given in Specifications. All springs must be of equal length, the maximum permissible difference in their free lengths being 1.0 mm (0.04 in).

8   When the clutch outer drum, bearing sleeve and centre are refitted to the input shaft with all the thrust washers, the clutch centre should be able to spin freely and independently of the outer drum when its retaining nut is fastened to the specified torque setting. Also, the outer drum should be able to spin freely on the input shaft.

9   If some stiffness is encountered, check that all components (particularly the thrust washers between the outer drum and centre) are correctly installed then set up a dial gauge to measure the outer drum end-float along the input shaft, and compare the reading obtained with those given in Specifications. If the end-float is too little or too much it can be adjusted by adding (or removing) shims of different thickness between the outer drum and centre.

10   Dismantle the assembly and measure (TY125 and TY175 only) the clutch outer drum centre bush, the input shaft and the bearing sleeve, renewing any component that is found to be excessively worn (see Specifications). Measurements are not provided for TY50 and TY80 models; each component should be checked for obvious signs of excessive wear and renewed if necessary. Finally, on all models, check the push rods for straightness by rolling on a sheet of plate glass or similar flat surface using feeler gauges to measure any distortion; this should not exceed 0.5 mm (0.02 in).

11   Check the components of the release mechanism for wear or damage and renew as necessary. Grease thoroughly on reassembly.

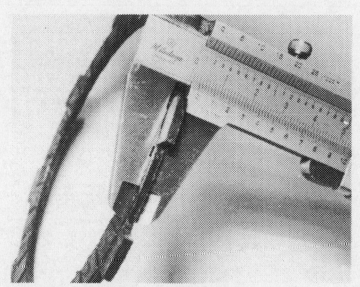

23.2 Measuring clutch friction plate thickness

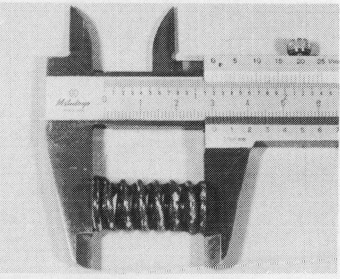

23.7 Measuring clutch spring free length

24 Examination and renovation: kickstart mechanism

1   The kickstart mechanism is a robust assembly and should not normally require attention. Apart from obvious defects such as a broken return spring, the friction clip is the only component likely to cause problems if it becomes worn or weakened. The clip is intended to apply a known amount of drag on the kickstart pinion, causing the latter to run up its quick thread and into engagement when the kickstarter lever is operated.

2   The clip can be checked using a spring balance. Hook one end of the balance onto the looped end of the friction clip. Pull on the free end of the balance and note the reading at the point where pressure overcomes the clip's resistance. This should normally be 0.8 - 1.5 kg (1.76 - 3.31 lb). If the reading is higher or lower than this and the mechanism has been malfunctioning, renew the clip as a precaution. Do not attempt to adjust a worn clip by bending it.

3   Examine the kickstart pinion for wear or damage, remembering to check it in conjunction with the output shaft-mounted idler pinion. In view of the fact that these components are not subject to continuous use, a significant amount of wear or damage is unlikely to be found.

25 Examination and renovation: gearbox components

1   Give the gearbox components a close visual inspection for signs of wear or damage such as broken or chipped teeth, worn dogs, damaged or worn splines and bent selectors. Replace any parts found unserviceable because they cannot be reclaimed in a satisfactory manner.

2   The gearbox shafts are unlikely to sustain damage unless the lubricating oil has been run low or the engine has seized and placed an unusually high loading on the gearbox. Check the surfaces of the shafts, especially where a pinion turns on them, and renew the shafts if they are scored or have picked up. The shafts can be checked for trueness by setting them up in V-blocks and measuring any bending with a dial gauge. The procedure for dismantling and rebuilding the gearbox shaft assemblies is given in the following Section.

3   Examine the gear selector claw assembly noting that worn or rounded ends on the claw can lead to imprecise gear selection. The springs in the selector mechanism and the detent or stopper arm should be unbroken and not distorted or bent in any way.

4   Examine the selector forks carefully, ensuring that there is no sign of scoring on the bearing surface of either of their claw ends, their bores or their selector drum guide pins. Check for any signs of cracking around the edges of the bores or at the base of the fork arms.

5   Check each selector fork shaft for straightness by rolling it on a sheet of plate glass and checking for any clearance between the shaft and the glass with feeler gauges. A bent shaft will cause difficulty in selecting gears. There should be no sign of any scoring on the bearing surface of the shaft or any discernible play between each shaft and its selector fork(s).

6   The tracks in the selector drum should not show signs of undue wear or damage. Check also that the selector drum bearing surfaces are unworn, renew the drum if it is worn or damaged.

7   Note that certain pinions have a bush fitted within their centres. If any one of these bushes appears to be over-worn or in any way damaged, then the pinion should be returned to a Yamaha service agent, who will be able to advise which course of action to take as to its renewal.

8   Finally, carefully inspect the splines of both shafts and pinions for any signs of wear, hairline cracks or breaking down of the hardened surface finish. If any one of these defects is apparent, then the offending component must be renewed. It should be noted that damage and wear rarely occur in a gearbox which has been properly used and correctly lubricated, unless a very high mileage has been covered.

9   Clean the gearbox sprocket thoroughly and examine it closely, paying particular attention to the condition of the teeth. The sprocket should be renewed if the teeth are hooked, chipped, broken or badly worn. It is considered bad practice to renew one sprocket on its own; both drive sprockets should be renewed as a pair, preferably with a new final drive chain. Examine the splined centre of the sprocket for signs of wear. If any wear is found, renew the sprocket as slight wear between the sprocket and shaft will rapidly increase due to the torsional forces involved. Remember that as the output shaft will probably wear in unison with the sprocket, it is therefore necessary to carry out a close inspection of the shaft splines.

10 Carefully examine the gear selector components. Any obvious signs of damage, such as cracks, will mean that the part concerned must be renewed. Check that the springs are not weak or damaged and examine all points of contact, eg gearchange shaft rollers and selector claw arm, for signs of excessive wear. If doubt arises about the condition of any component it must be compared with a new part to assess the amount of wear that has taken place, and renewed if found to be damaged or excessively worn. Do not forget to check the pins set in the selector drum; these must be renewed if bent or worn.

11 On TY125 and TY175 models, check with particular care the linkage on the left-hand side of the engine. If the seal fails, dirt and water get in to accelerate wear and cause imprecise gear selection through the excessive free play that results. If it is worn, the bush in the cover can be pressed out and a new one fabricated, this is a task for a light engineering company as Yamaha only supply complete covers. All other components should be thoroughly cleaned, and renewed if worn, to eliminate as much free play as possible. Renew the cover seal as a matter of course.

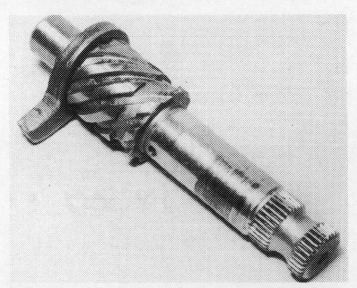

24.1 Examine kickstart assembly for signs of wear or damage, especially on curved splines

24.2 Friction clip resistance can be measured to assess clip condition

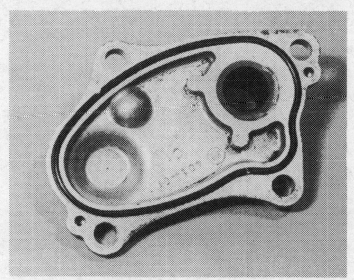

25.11 TY125 and 175 – check gearchange components on left-hand side of engine – particularly cover seals and bush

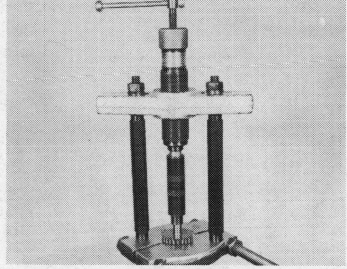

26.3 TY50 – input shaft 5th gear pinion must be removed using hydraulic press – measure shaft before pinion is removed

26 Gearbox shafts: dismantling and reassembly

Dismantling – general

1   The gearbox clusters should not be disturbed needlessly, and need only be stripped when careful examination of the whole assembly fails to reveal the cause of a problem, or where obvious damage, such as stripped or chipped teeth is discovered.

2   The input and output shaft components should be kept separate to avoid confusion during reassembly. Using circlip pliers, remove the circlip and plain washer which retain each part. As each item is removed, place it in order on a clean surface so that the reassembly sequence is obvious and the risk of parts being fitted the wrong way round or in the wrong sequence is avoided. Care should be exercised when removing circlips to avoid straining or bending them excessively. The clips must be opened just sufficiently to allow them to be slid off the shaft. Note that a loose or distorted circlip might fail in service, and any dubious items must be renewed as a precautionary measure. The same applies to worn or distorted thrust washers.

3   If dismantling of the input shaft assembly proves to be necessary, on TY50 models only, the 5th gear pinion will have to be pressed from position by using an hydraulic press; no other method of removal is possible. As it is unlikely that this type of tool will be readily available, it is recommended that the complete shaft assembly be returned to a Yamaha service agent who will be able to remove the pinion, renew any worn or damaged components and return the shaft assembly complete.

4   If an hydraulic press is available and it is decided to attempt removal of the 5th gear pinion from the input shaft, it is very important to realise fully the dangers involved when using such a tool. Both the tool and the shaft assembly must be set up so that there is no danger of either item slipping. The tool must be correctly assembled in accordance with the maker's instructions as the force exerted by the tool is considerable and perfectly capable of stripping any threads from holding studs or inflicting other damage upon itself and the shaft. Always wear proper eye protection in case a component should fail and shatter before it becomes free, as may happen if the component is flawed. **Before** removing the 5th gear pinion, measure carefully the distance between the outer faces of the 1st and 5th gear pinions, then measure with feeler gauges the clearance between the 5th and 2nd gear pinions. Be very careful; it is essential that these measurements are made before the shaft is disturbed, so that it can be reassembled correctly.

5   The accompanying illustration shows how both clusters of the gearbox are assembled on their respective shafts. It is imperative that the gear clusters, including the thrust washers, are assembled in exactly the correct sequence, otherwise constant gear selection problems will occur. In order to eliminate the risk of misplacement, make rough sketches as the clusters are dismantled. Also strip and rebuild as soon as possible to reduce any confusion which might occur at a later date.

Reassembly – general

6   Having checked and renewed the gearbox components as required (see Section 25), reassemble each shaft, referring to the accompanying line drawing and photographs for guidance. Note that the manufacturer specifies a particular way in which the circlips are to be fitted. If a non-wire type circlip is examined closely, it can be seen that one surface has rounded edges and the other has sharply-cut square edges, this feature being a by-product of any stamped component such as a circlip or thrust washer. The manufacturer specifies that each circlip must be fitted with the sharp-edged surface facing away from any thrust received by that circlip. In the case of the gearbox shaft circlips this means that the rounded surface of any circlip must face towards the gear pinion that it secures. Furthermore, when a circlip is fitted to a splined shaft, the circlip ears must be positioned in the middle of one of the splines. These two simple precautions are specified to ensure that each circlip is as secure as is possible on its shaft and is best able to carry out its task. Ensure that the bearing surfaces of each component are liberally oiled before fitting.

7   If problems arise in identifying the various gear pinions which cannot be resolved by reference to the accompanying photographs and illustrations, the number of teeth on each pinion will identify them. Count the number of teeth on the pinion and compare this figure with that given in the Specifications Section of this Chapter, remembering that the output shaft pinions are listed first, followed by those on the input shaft. The problem of identification of the various components should not arise, however, if the instructions given in paragraph 2 of this Section are followed carefully.

8   On TY50 models only, when reassembling the input shaft fit all other components then thoroughly degrease both the shaft left-hand end and the 5th gear pinion. Apply a locking compound, such as Loctite Bearing Fit, to the internal surface of the pinion and push it as far as possible over the shaft end and down the length of the shaft. The pinion must be fitted so that the protruding shoulder faces to the right. A means must now be found of pressing the pinion on to the shaft to a precise position. While this can be achieved, using a hammer and a long tubular drift, the method employed in practice was as follows. A vice with a jaw opening of at least 8 inches must be found. Place a soft alloy or wooden cover over one of the vice jaws and place the threaded right-hand end of the shaft against this cover. Slip a socket or tubular drift of suitable length and diameter over the left-hand end of the shaft and position this against the other vice jaw. With the socket or tubular drift bearing only on the surface on the pinion, tighten the vice until the outer faces of the 1st and 5th gear pinions are precisely the previously measured distance apart; this can be confirmed by measuring the

clearance between the 5th and 2nd gear pinions. Check the measurement frequently as the vice is tightened, using a micrometer or vernier gauge for absolute accuracy.

9   Spin the 2nd gear pinion by hand; it is essential that it is free to rotate easily. If the pinion becomes locked, whether by contact with the 5th gear pinion or by the action of the locking compound, the 5th gear pinion must be removed again for all components to be cleaned, checked, and refitted correctly.

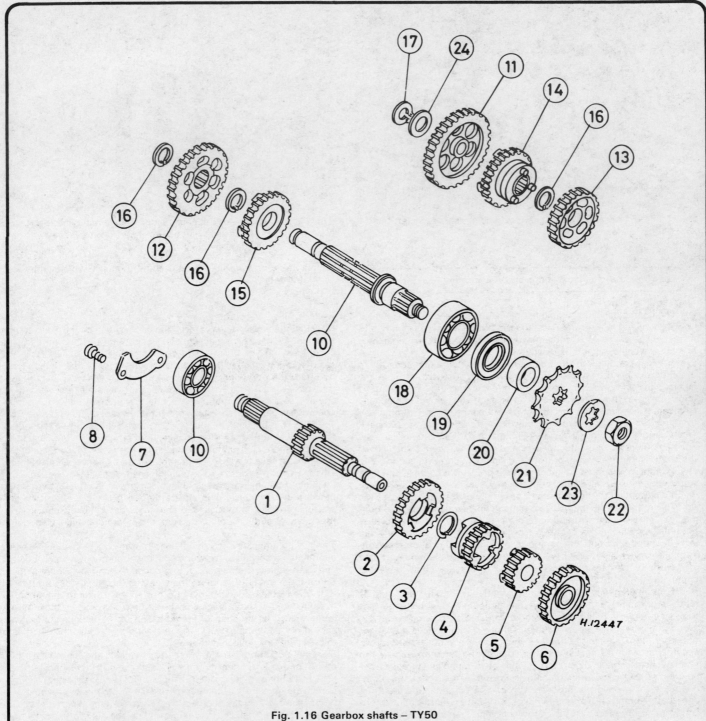

Fig. 1.16 Gearbox shafts – TY50

1 Input shaft	7 Bearing retainer	13 Output shaft 3rd gear	19 Oil seal
2 Input shaft 4th gear	8 Screw – 2 off	14 Output shaft 4th gear	20 Spacer
3 Circlip	9 Bearing	15 Output shaft 5th gear	21 Gearbox sprocket
4 Input shaft 3rd gear	10 Output shaft	16 Circlip – 3 off	22 Nut
5 Input shaft 2nd gear	11 Output shaft 1st gear	17 Circlip	23 Lock washer
6 Input shaft 5th gear	12 Output shaft 2nd gear	18 Bearing	24 Thrust washer

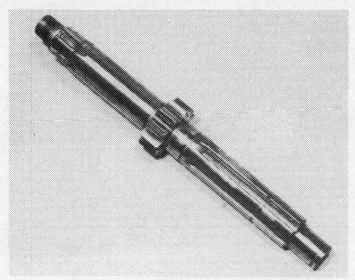

26.6a Gearbox shaft reassembly TY175 – input shaft identififed by integral 1st gear

26.6b Fit 5th gear pinion as shown ...

26.6c ... followed by a plain thrust washer

26.6d Secure the pinion with a circlip

26.6e Fit double 3rd/4th gear as shown ...

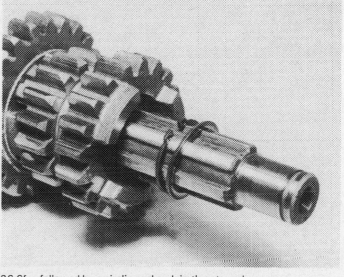

26.6f ... followed by a circlip and a plain thrust washer

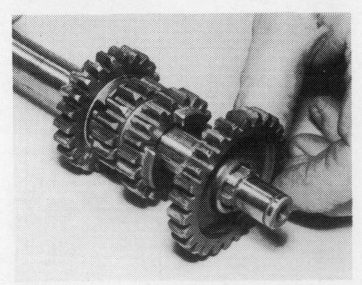

26.6g 6th gear pinion recessed face points to left ...

26.6h ... and is followed by 2nd gear pinion

26.6i Lubricate bearing thoroughly before refitting

26.6j Refit circlip to retain all components

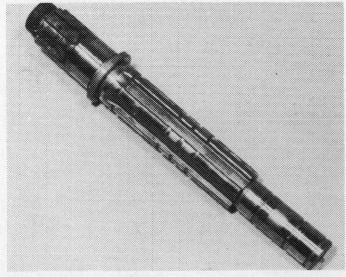

26.7a Take the bare output shaft ...

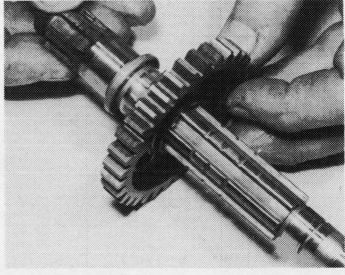

26.7b ... and fit the 2nd gear pinion, as shown ...

26.7c ... followed by a plain thrust washer ...

26.7d ... and a circlip

26.7e Fit the 6th gear pinion ...

26.7f ... followed by a circlip

26.7g The 4th gear pinion (27T) is fitted next ...

26.7h ... followed by a circlip ...

26.7i ... and the 3rd gear pinion (30T)

26.7j Fit a splined thrust washer ...

26.7k ... followed by a circlip ...

26.7l ... and the 5th gear pinion

26.7m Lastly fit the 1st gear pinion as shown ...

26.7n ... followed by the large thrust washer ...

26.7o ... and a circlip

26.8 Fitting the input shaft 5th gear pinion – TY50 – note careful use of vice

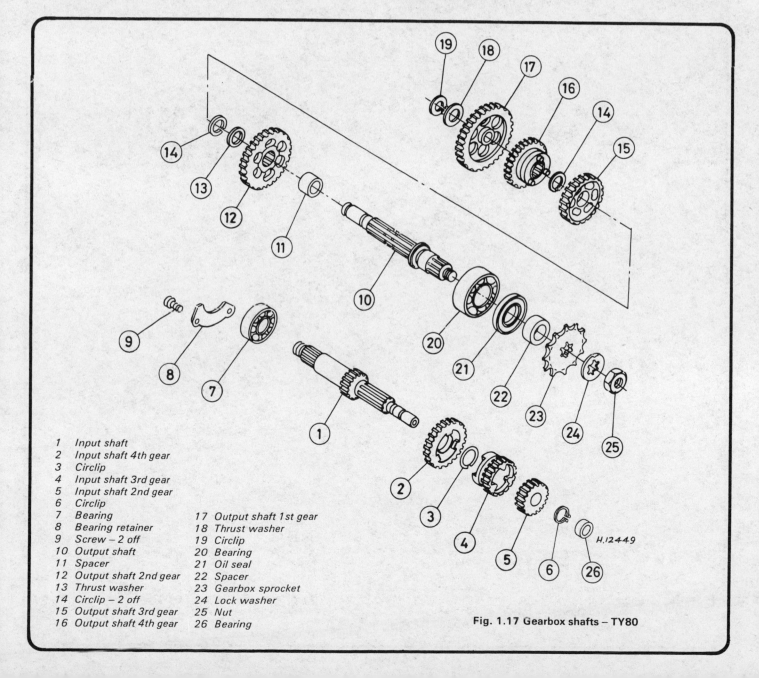

1 Input shaft
2 Input shaft 4th gear
3 Circlip
4 Input shaft 3rd gear
5 Input shaft 2nd gear
6 Circlip
7 Bearing
8 Bearing retainer
9 Screw – 2 off
10 Output shaft
11 Spacer
12 Output shaft 2nd gear
13 Thrust washer
14 Circlip – 2 off
15 Output shaft 3rd gear
16 Output shaft 4th gear

17 Output shaft 1st gear
18 Thrust washer
19 Circlip
20 Bearing
21 Oil seal
22 Spacer
23 Gearbox sprocket
24 Lock washer
25 Nut
26 Bearing

H.12449

Fig. 1.17 Gearbox shafts – TY80

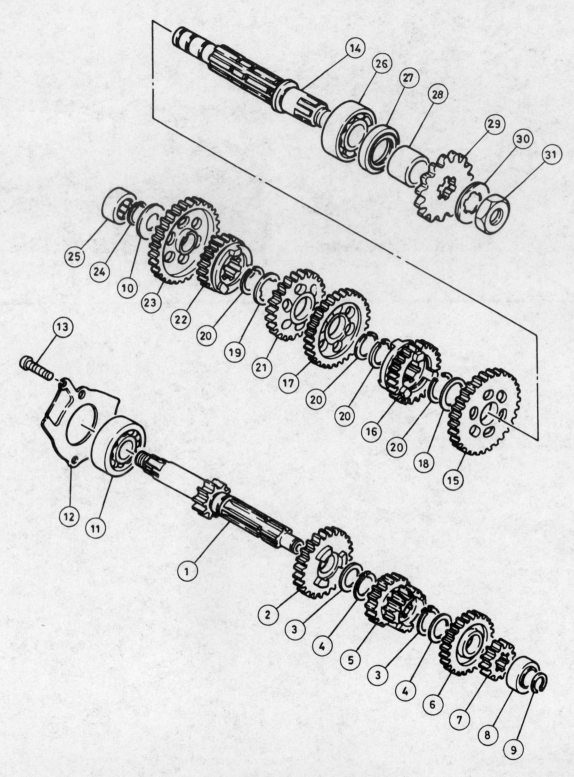

Fig. 1.18 Gearbox shafts – TY125

1	Input shaft	9	Circlip	17	3rd gear pinion	25	Bearing
2	6th gear pinion	10	Thrust washer	18	Thrust washer	26	Bearing
3	Thrust washer	11	Bearing	19	Splined thrust washer	27	Oil seal
4	Circlip	12	Bearing retainer	20	Circlip	28	Spacer
5	3rd/4th gear pinion	13	Screw	21	4th gear pinion	29	Sprocket
6	5th gear pinion	14	Output shaft	22	6th gear pinion	30	Lock washer
7	2nd gear pinion	15	2nd gear pinion	23	1st gear pinion	31	Nut
8	Bearing	16	5th gear pinion	24	Circlip		

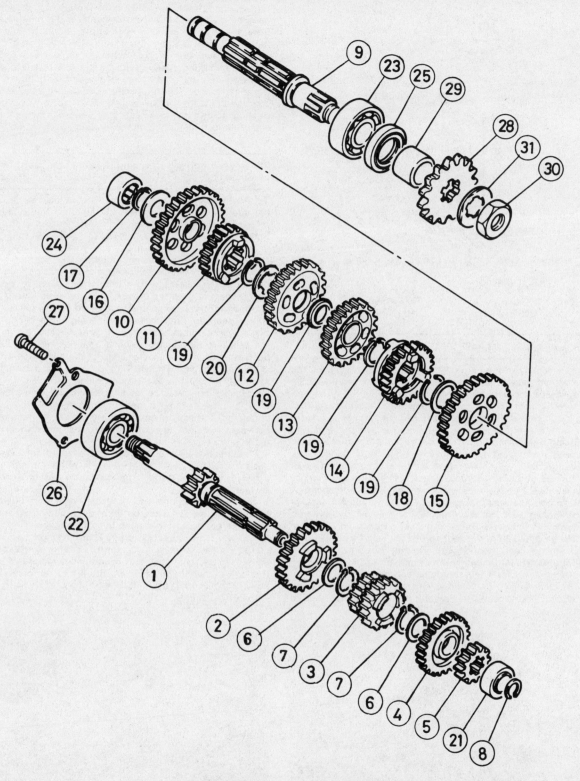

Fig. 1.19 Gearbox shafts – TY175

1 Input shaft	9 Output shaft	17 Circlip	25 Oil seal
2 Input shaft 5th gear	10 Output shaft 1st gear	18 Thrust washer	26 Bearing retainer
3 Input shaft 3rd/4th gear	11 Output shaft 5th gear	19 Circlip – 4 off	27 Screw – 2 off
4 Input shaft 6th gear	12 Output shaft 3rd gear	20 Splined thrust washer	28 Gearbox sprocket
5 Input shaft 2nd gear	13 Output shaft 4th gear	21 Bearing	29 Spacer
6 Thrust washer – 2 off	14 Output shaft 6th gear	22 Bearing	30 Nut
7 Circlip – 2 off	15 Output shaft 2nd gear	23 Bearing	31 Lock washer
8 Circlip	16 Shim	24 Bearing	

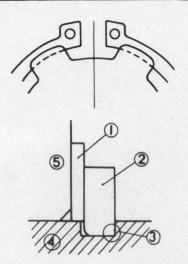

Fig. 1.20 Correct fitting of a circlip to a splined shaft

1 Thrust washer   4 Shaft
2 Circlip   5 Gear pinion
3 Square edge

27 Engine reassembly: general

1 Before reassembly of the engine/gearbox unit is commenced, the various component parts should be cleaned thoroughly and placed on a sheet of clean paper, close to the working area.

2 Make sure all traces of old gaskets have been removed and that the mating surfaces are clean and undamaged. Great care should be taken when removing old gasket compound not to damage the mating surface. Most gasket compounds can be softened using a suitable solvent such as methylated spirits, acetone or cellulose thinner. The type of solvent required will depend on the type of compound used. Gasket compound of the non-hardening type can be removed using a soft brass-wire brush of the type used for cleaning suede shoes. A considerable amount of scrubbing can take place without fear of harming the mating surfaces. Some difficulty may be encountered when attempting to remove gaskets of the self-vulcanising type. The gasket should be pared from the mating surface using a scalpel or a small chisel with a finely honed edge. Do not, however, resort to scraping with a sharp instrument unless necessary.

3 Gather together all the necessary tools and have available an oil can filled with clean engine oil. Make sure that all new gaskets and oil seals are to hand, also all replacement parts required. Nothing is more frustrating than having to stop in the middle of a reassembly sequence because a vital gasket or replacement part has been overlooked. As a general rule each moving engine component should be lubricated thoroughly as it is fitted into position.

4 Make sure that the reassembly area is clean and that there is adequate working space. Refer to the torque and clearance setting wherever they are given. Many of the smaller bolts are easily sheared if overtightened. Always use the correct size screwdriver bit for the cross-head screws and never an ordinary screwdriver or punch. If the existing screws show evidence of maltreatment in the past, it is advisable to renew them as a complete set.

28 Reassembling the engine/gearbox unit: preparing the crankcases

1 At this stage the crankcase castings should be clean and dry with any damage, such as worn threads, repaired. If any bearings are to be refitted, the crankcase casting must be heated first as described in Section 14.

2 Place the heated casting on a wooden surface, fully supported around the bearing housing. Position the bearing on the casting, ensuring that it is absolutely square to its housing then tap it fully into place using a hammer and a tubular drift such as a socket spanner which bears only on the bearing outer race. Be careful to ensure that the bearing is kept absolutely square to its housing at all times.

3 Oil seals are fitted into a cold casing in a similar manner. Apply a thin smear of grease to the seal circumference to aid the task, then tap the seal into its housing using a hammer and a tubular drift which bears only on the hard outer edge of the seal, thus avoiding any risk of the seal's being distorted. Tap each seal into place until its flat outer surface is just flush with the surrounding crankcase. Oil seals are fitted with the spring-loaded lip towards the liquid (or gas) being retained, ie with the manufacturer's marks or numbers facing outwards. Where double-lipped seals are employed, eg right-hand main bearing, use the marks or numbers to position each seal correctly if no notes were made on removal.

4 Where retaining plates are employed to secure bearings or oil seals, thoroughly degrease the threads of the mounting screws, apply a few drops of thread locking compound to them, and tighten them securely.

5 When all bearings and oil seals have been fitted and secured, lightly lubricate the bearings with clean engine oil and apply a thin smear of grease to the sealing lips of each seal.

6 Support the appropriate crankcase half on two wooden blocks placed on the work surface; there must be sufficient clearance to permit the crankshaft and gearbox components to be fitted. Remember that these are to be fitted into the crankcase right-hand half on TY50 and TY80 models, and into the left-hand half on TY125 and TY175 models.

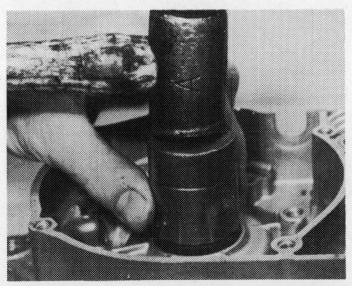

28.3 Oil seals and bearings are refitted as shown

28.4 Use thread locking compound on retaining plate screws

29 Reassembling the engine/gearbox unit: refitting the crankshaft and the gearbox components

1   Refit the appropriate nut to protect the crankshaft threaded end and insert the crankshaft as far as possible into its main bearing, using a smear of oil to ease the task. On TY125 and TY175 models, do not forget the thrust washer on the crankshaft left-hand end. Align the connecting rod with the crankcase mouth, check that the crankshaft is square to the crankcase and support the flywheels at a point opposite the crankpin to prevent distortion while the crankshaft is driven home with a few firm blows from a soft-faced mallet. Do not risk damaging the crankshaft by using excessive force; if undue difficulty is encountered, take the assembly to a Yamaha Service Agent for the crankshaft to be drawn into place using the correct service tool.

2   When the crankshaft is fitted, remove the protecting nut and check that the crankshaft revolves easily with no trace of distortion. On TY125 and 175 models fit a new sealing O-ring to the right-hand mainshaft groove and smear it with grease.

3   Fit the gear clusters together, ensuring that all pinions are correctly meshed, and assemble the selector forks and drum. Refer to the accompanying line drawings (and photographs, if applicable) to identify correctly each fork. If no notes were made on dismantling, the forks themselves may be marked by the manufacturer. On the machine featured in the photographs each fork has a number ('1', '2', or '3') cast in its right-hand face (ie uppermost when the unit is dismantled on the bench). Number '1' fork fits on the output shaft 5th gear pinion, number '3' fork fits on the output shaft 4th gear pinion and number '2' fork fits on the input shaft 3rd/4th gear pinion. A similar system of marking may be found on other models. Check that all fork guide pins are securely refitted. On TY50 and TY80 models, always use a new split-pin to secure the large selector fork guide pin in the selector drum centre track and ensure that it is neatly trimmed so that it cannot foul any other component when installed; also, do not forget to refit the selector cam on the drum left-hand end.

4   With the gearbox components correctly assembled and held as a single unit, insert them into the crankcase and press into place; a few taps with a soft-faced mallet may be necessary, but be very careful not to dislodge or damage any component.

5   On TY125 and TY175 models position the single selector fork on its pinion groove and selector drum track, then press in the fork shaft. On all models select neutral and check that both shafts rotate easily, then lubricate all bearing surfaces thoroughly.

29.1 Refitting the crankshaft into the crankcase left-hand half – TY125 and 175 – do not omit thrust washer

29.2 Fit a new O-ring to crankshaft and smear with grease for protection – TY125 and 175

29.3a TY50 – arrangement of selector forks on output shaft ...

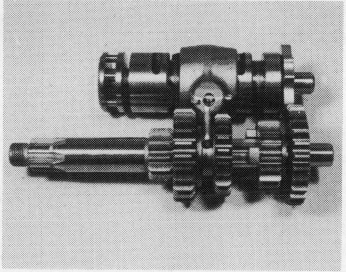

29.3b ... and input shaft – note split pin retaining fork guide pin

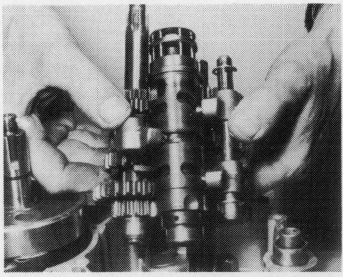

29.4 Assemble selector forks and gear clusters to be fitted as a single unit – note selector drum notch in neutral position

29.5 Fit shorter selector fork shaft as shown – TY125 and 175

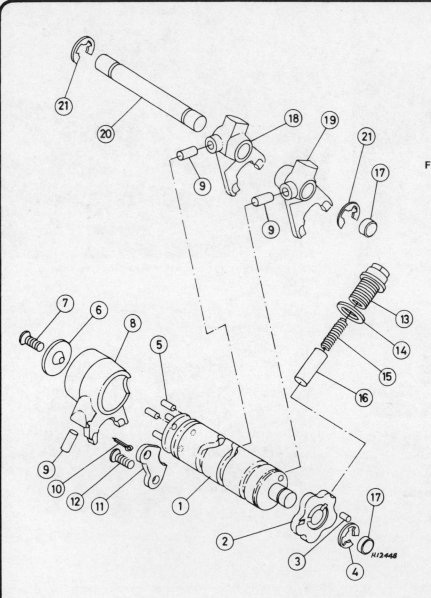

Fig. 1.21 Selector drum and forks – TY50 and 80

1   Selector drum
2   Selector cam
3   Dowel pin
4   Circlip
5   Selector pins
6   Drum end cover
7   Screw
8   Selector fork 1
9   Guide pin – 3 off TY50, 2 off TY80
10 Split pin
11 Guide plate
12 Screw – 2 off
13 Detent plunger cap
14 Sealing washer
15 Spring
16 Detent plunger
17 Plug
18 Selector fork 2
19 Selector fork 3 – TY50 only
20 Selector fork shaft
21 Circlip – 2 off

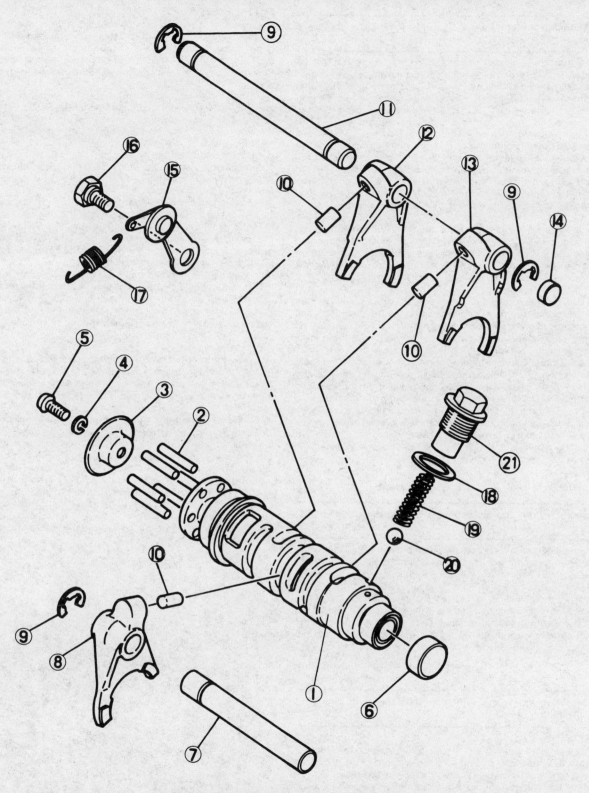

Fig. 1.22 Selector drum and forks – TY125 and 175

1	Selector drum	8	Selector fork 2	15	Stopper lever assembly
2	Dowel pin – 6 off	9	Circlip – 3 off	16	Bolt
3	End cover	10	Guide pin	17	Spring
4	Spring washer	11	Shaft	18	Washer
5	Screw	12	Selector fork 3	19	Spring
6	Plug	13	Selector fork 1	20	Steel ball
7	Shaft	14	Plug	21	Detent plunger cap

30 Reassembling the engine/gearbox unit: joining the crankcase halves

1   Apply a thin film of sealing compound to the gasket surface of the lower crankcase half, then press the two locating dowels firmly into their recesses in the crankcase mating surface. Make a final check that all components are in position and that all bearings and bearing surfaces are lubricated. On TY125 and TY175 models, do not forget to refit the dowels and O-rings at the engine mounting points.

2   Lower the upper crankcase half into position, using firm hand pressure only to push it home. It may be necessary to give a few gentle taps with a soft-faced mallet to drive the casing fully into place. Do not use excessive force, instead be careful to check that all shafts and dowels are correctly fitted and accurately aligned, and that the crankcase halves are exactly square to each other. If necessary, pull away the upper crankcase half to rectify the problem before starting again.

3   When the two halves have joined correctly and without strain, refit the crankcase retaining screws using the cardboard template to position each screw correctly. Working in a diagonal sequence from the centre outwards, progressively tighten the screws until all are securely and evenly fastened; the recommended torque setting is 0.9 − 1.1 kgf m (6.5 − 8.0 lbf ft).

4   Wipe away any excess sealing compound from around the joint area, then check the free running and operation of the crankshaft and gearbox components. If a particular shaft is stiff to rotate, a smart tap on each end using a soft-faced mallet, will centralise the shaft in its bearing. If this does not work, or if any other problem is encountered, the crankcases must be separated again to find and rectify the fault. Pack clean rag into the crankcase mouth to prevent the entry of dirt, then refit the drain plug, if removed, tightening it to the specified torque setting. On TY50 and TY80 models apply thread locking compound to the threads of its retaining screws and refit the selector drum guide plate, tightening the screws securely with an impact driver. On all models refit the gearbox sprocket spacer to the output shaft left-hand end, using a generous smear of grease to ensure that there is no damage to the oil seal.

31 Reassembling the engine/gearbox unit: refitting the flywheel generator

1   Relocate the stator in its previously noted position using the wiring as a guide and tighten its retaining screws. Clip the electrical leads to the crankcase and reconnect the neutral indicator switch (where fitted).

2   Degrease the rotor and crankshaft mating surfaces. Insert the Woodruff key into the crankshaft and push the rotor over it. Gently tap the rotor centre with a soft-faced hammer to seat it and fit the plate washer, spring washer and nut. Lock the crankshaft and tighten the nut to the specified torque loading.

32 Reassembling the engine/gearbox unit: refitting the gear selector external components

TY50, TY80

1   Insert the selector pins into the selector drum, apply thread locking compound to the threads of its retaining screw and refit the drum end cover, using an impact driver to tighten the screw securely.

2   Lubricate the plunger and spring of the detent assembly, check that they slide easily in the cap, and refit the assembly with a new sealing washer.

3   Smear grease over the gearchange shaft splines to protect the oil seal and insert the shaft fully into the crankcase. Place the plain washer over its left-hand end, then refit the retaining circlip. Fit the roller to its right-hand end.

4   Fit the return spring to the selector claw arm assembly; it is installed in the same way as the TY175 component shown in the accompanying photograph. Place the assembly on its pivot, engaging the return spring ends on each side of the stop and the claw arm on the selector pins. Refit the circlip to the pivot.

TY125, TY175

5   Insert the selector pins into the selector drum, apply thread locking compound to the threads of its retaining screw and refit the drum end cover, using an impact driver to tighten the screw securely.

6   Lubricate the ball and spring of the detent assembly, check that they move easily in the cap, and refit the assembly with a new sealing washer.

7   Place the detent roller arm in position on the crankcase, apply a few drops of thread locking compound to its threads and refit the pivot bolt; tighten it securely but take care not to overtighten it. Check that the roller arm can move easily and hook its spring on to the bearing retaining plate.

8   Fit the return spring to the gearchange shaft as shown in the accompanying photograph and insert the shaft into the crankcase, engaging the return spring ends on each side of the stop and the claw arm on the selector pins.

9   Working on the engine left-hand side, fit a thrust washer on each side of the gearchange pedal shaft, smear it liberally with grease and fit it into its recess. Fit the roller to its post. Place a thrust washer over the gearchange shaft splines and refit the linkage arm over the gearchange shaft end and the pedal shaft roller so that the centre of the roller is in line with the two shaft centres (see accompanying photographs). Fit the circlip to retain the assembly.

10   Pack the whole cavity full of grease to provide protection against dirt and corrosion; the transmission oil does not reach this cavity. Press the two dowel pins into their holes and refit the linkage cover with a new sealing O-ring. Tighten the four screws securely.

All models

11   With the mechanism fully refitted, select second gear and check that the claw arm stops are exactly equidistant from the two nearest selector pins as shown in the accompanying illustrations. This is essential to ensure that the mechanism has exactly the correct amount of movement for positive gear selection.

12   Adjustment is made by flattening back the lockwasher (TY125, TY175 only) slackening its locknut and rotating the return spring stop which is actually an eccentric adjusting screw. When both clearances are the same as shown, hold the screw and tighten the locknut securely. Secure the locknut by bending up against one of its flats an unused portion of the lock washer (TY125, TY175 only). On TY125 and TY175 models if excessive adjustment is required, do not forget to check that the linkage on the engine left-hand side is still correctly aligned.

13   Check that all gears can be selected with relative ease and return to the neutral position.

30.1a Ensure all components are refitted and lubricated before joining crankcases – note locating dowels (arrowed)

30.1b Do not omit dowel and O-rings at engine mounting – TY125 and 175

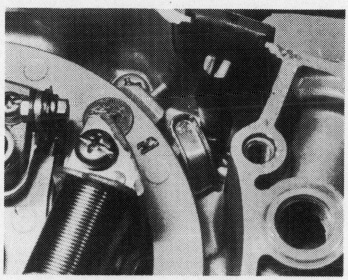

31.1 Ensure wiring is correctly routed and secured on refitting stator plate

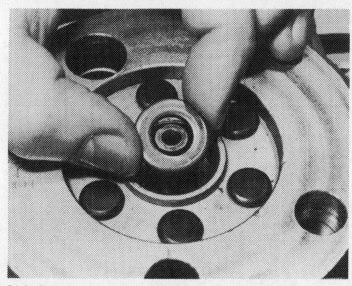

31.2a Refit the generator rotor, followed by the plain washer ...

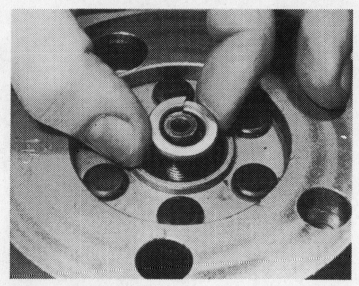

31.2b ... and spring washer

31.2c Lock crankshaft and tighten retaining nut to specified torque setting

32.5a Refit the selector pins into the selector drum

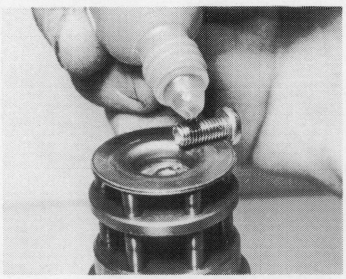

32.5b Apply thread locking compound to the screw on refitting the drum end cover

32.6 Assemble detent ball and spring into cap and retain with grease on refitting

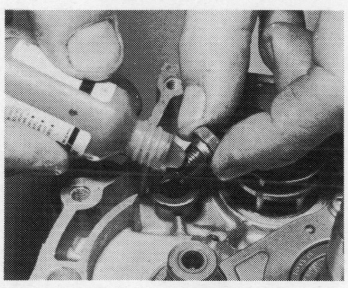

32.7a Apply thread locking compound to detent roller arm pivot bolt ...

32.7b ... check arm is free to move, then hook spring to retaining plate

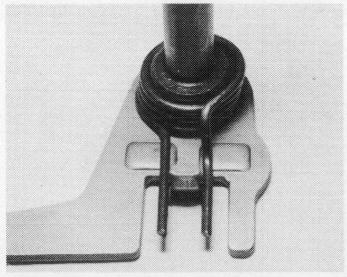

32.8 Fit gearchange shaft return spring free ends as shown

32.9a Position thrust washers over gearchange shaft and pedal shaft bore as shown ...

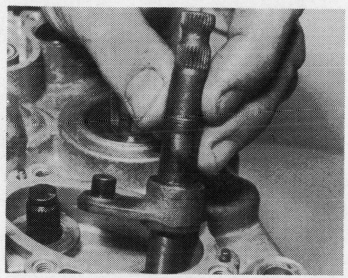

32.9b ... then grease and refit pedal shaft

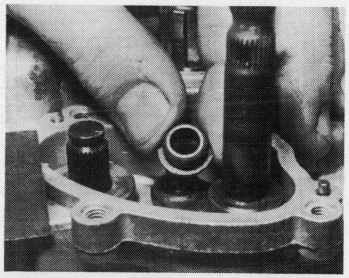

32.9c Fit roller over pedal shaft post, as shown ...

32.9d ... and refit second pedal shaft thrust washer

32.9e Fit linkage arm, aligning it as shown ...

32.9f ... and secure by refitting circlip

32.10 Pack cavity full of grease and use new sealing O-ring – note two small locating dowels (arrowed)

32.11 Check that claw arm is correctly aligned with selector pins – see text

32.12 If necessary, adjustment is made at return spring stop

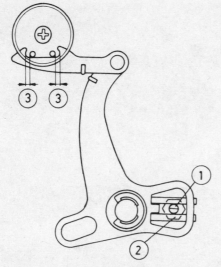

Fig. 1.23 Selector arm centralising adjustment – TY50 and 80

1   Adjusting screw   3   Claw arm to pin clearance
2   Locknut

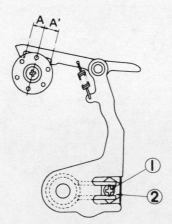

Fig. 1.24 Selector arm centralising adjustment – TY125 and 175

1   Adjusting screw   A   Claw arm to pin clearance
2   Locknut

33 Reassembling the engine/gearbox unit: refitting the kickstart assembly

1   Refit to the output shaft right-hand end the circlip or plain washer (as appropriate), followed by the wave or plain washer and the idler pinion. On TY50 and TY80 models the pinion's deeply recessed face **must** face outwards, while on TY125 and TY175 models its chamfered outer edge **must** face outwards. Refit the second plain washer and the circlip.

2   If dismantled, rebuild the kickstart shaft assembly with reference to Fig. 1.8 and 1.9 and (on TY125 and TY175 models only) the series of photographs. Insert it into the casing aligning the shaft stop with the lug on the crankcase wall and the friction clip looped end with its recess. Check that the mechanism is operating correctly.

3   Bring the long hooked end of the return spring around so that it can be engaged on the long stop dowel protruding from the crankcase wall.

33.1a Refit the circlip (or plain washer) to the output shaft right-hand end, followed by ...

33.1b ... the idler pinion and thrust washers – secure with a circlip

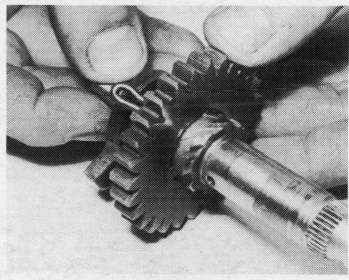

33.2a Fit kickstart pinion and friction clip as shown ...

33.2b ... and insert collets into shaft groove ...

33.2c ... to be secured by the large circlip

33.2d Refit large plain washer against shaft shoulder

33.2e Return spring inner end fits into hole in shaft ...

33.2f ... slot in spring guide engages on spring inner end

33.2g Ensure shaft stop and friction clip engage in correct crankcase recesses (arrowed)

33.3 Engage kickstart return spring on spring stop

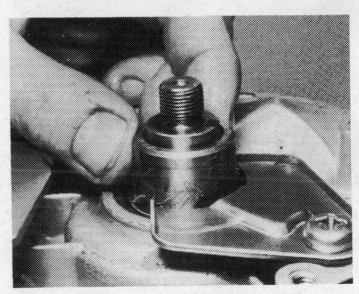

34.1a Grease seal lips and spacer to aid refitting

34 Reassembling the engine/gearbox unit: refitting the clutch and primary drive pinion

1   Grease the lips of the crankshaft right-hand oil seal and refit the spacer to the crankshaft, taking care not to damage the seal, then refit the Woodruff key followed by the primary drive pinion with its recessed face outwards. Where the retaining nut lock washer is of the Belleville (conical) type, it is refitted with its convex surface outwards. Refit the nut, lock the crankshaft and tighten the nut to the torque setting specified.

2   On TY125 and TY175 models refit the return spring and plain washer to the release shaft, smear it with grease and insert it into the crankcase. Check that the return spring is correctly located, then refit the adjusting screw with a new sealing O-ring and screw it fully in. Lift the release shaft up and down until the screw eccentric pin engages in the shaft groove; the shaft will move slightly up and down as the screw rotates. Do not apply excessive force to the screw at any time or the eccentric pin will shear off. Refit the locknut loosely.

3   Place the large thrust washer over the input shaft end, followed by the bearing sleeve and the clutch outer drum, the smaller thrust washer(s) and the clutch centre. Fit the lockwasher and retaining nut, lock the clutch centre and tighten the nut to the specified torque setting. If a tab-type lock washer is fitted, secure the nut by bending up against one of its flats an unused portion of the lock washer; if a Belleville (conical) washer is used, ensure it is positioned as described above.

4   Insert the plain pushrod into the input shaft noting that its grooved end must be in the centre of the shaft on TY50 and TY80 models. This is followed by the steel ball and the mushroom-headed pushrod.

5   Starting with a friction plate followed by a plain plate, refit the clutch friction and plain plates alternately to finish with a friction plate. New friction plates should be coated heavily with engine oil. On TY50 and TY80 models, note that a rubber cushion ring is fitted around the clutch centre inside each friction plate. Also on these models, do not forget to refit the O-ring around the outer drum.

6   Fit the clutch pressure plate, ensuring it engages correctly on the clutch centre splines, then refit the springs and their bolts. Tighten the bolts progressively and in a diagonal sequence to a torque setting of 0.7 – 1.0 kgf m (5.0 – 7.0 lbf ft).

34.1b Do not omit Woodruff key

34.1c Lock washer is of Belleville type – convex surface outwards

34.1d Lock crankshaft to tighten retaining nut to specified torque setting

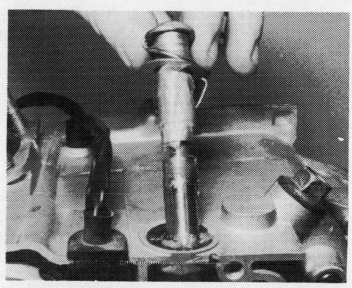

34.2a Fit return spring and washer to release shaft – grease before refitting

34.2b Refit clutch adjusting screw – do not apply excessive force

34.3a Place large thrust washer over input shaft end ...

34.3b ... followed by clutch bearing sleeve ...

34.3c ... and clutch outer drum

34.3d More than one thrust washer may be fitted between outer drum and clutch centre – see Section 23

34.3e Refit clutch centre – retaining nut lock washer will be of tab type ...

34.3f ... which is secured as shown – TY125 and 175 ...

34.3g ... or will be of Belleville type with convex surface outwards – TY50 and 80

34.3h Use fabricated tool to hold clutch centre while retaining nut is tightened

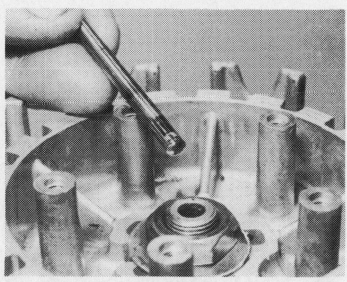

34.4a Insert plain pushrod into input shaft ...

34.4b ... followed by steel ball ...

34.4c ... and mushroom-headed pushrod

34.5a Refit clutch plates, alternating plain (steel) ...

34.5b ... and friction plates

34.6 Refit clutch springs and retaining bolts

35.1 Fit new cover gasket over locating dowels (arrowed)

35 Reassembling the engine/gearbox unit: refitting the right-hand crankcase cover

1 Check all components within the cover are properly fitted and lubricated. Grease the splined end of the kickstart shaft and the lip of the cover oil seal. If it was removed, check that the oil feed pipe sealing grommet is pressed fully into its cutout; this can be done with the cover fitted but it is much more difficult. Check that the feed pipes are correctly routed.
2 Degrease the cover and crankcase mating surfaces, press the locating dowels into the crankcase and locate the new gasket over them. Push the cover over the dowels, tapping it lightly with a soft-faced hammer to seat it properly and taking care to engage the pump drive and primary drive pinions.
3 Fit each screw into its previously noted position. Tighten the screws evenly whilst working in a diagonal sequence.

36 Reassembling the engine/gearbox unit: refitting the piston, cylinder barrel and head

1 The piston rings are refitted using the same technique as on removal. Do not forget to fit the expander to the bottom piston ring groove before refitting that ring. If neither surface is marked to show which is the top, plain piston rings can be fitted either way up when new; if the original rings are being refitted they must be installed the original way up using the wear marks for identification. Keystone rings have a slight inwards taper on their top surfaces; check carefully before refitting to ensure that the tapered surface is upwards. The Dykes rings fitted to the top groove on TY125 and TY175 models are fitted as shown in the accompanying illustration.
2 In all cases, rotate the rings so that their end gaps align with the locating peg in each ring groove to stop the rings rotating and catching in one of the ports. See the accompanying illustration.
3 Lubricate the big-end bearing and main bearings, bring the connecting rod to the top of its stroke, then pack the crankcase mouth with clean rag. Lubricate and fit the small-end bearing. With the arrow cast in the piston crown facing forward, place the piston over the rod and fit the gudgeon pin. If necessary, warm the piston to aid fitting.
4 Retain the gudgeon pin with **new** circlips. Check each clip is correctly located; if allowed to work loose it will cause serious damage.
5 Check the piston rings are still correctly fitted. Clean the barrel and crankcase mating surfaces and fit the new base gasket. Lubricate the piston rings and cylinder bore. Lower the barrel over its retaining studs and ease the piston into the bore, carefully squeezing the ring ends together. Do not use excessive force.
6 Remove the rag from the crankcase mouth and push the barrel down onto the crankcase. Clean the barrel to head mating surfaces and fit a new head gasket. Fit the head and its retaining nuts with washers. Tighten the nuts evenly and in a diagonal sequence to the specified torque setting.
7 Fit a new gasket to the intake port and refit the reed valve assembly, followed by a second gasket and the intake stub. Tighten the four retaining bolts evenly to a torque setting of 0.8 kgf m (6.0 lbf ft).
8 On TY50 and TY80 models, the oil feed pipe can be reconnected, but remember that the bleeding procedure must be carried out.
9 On all models, check its gap as described in Routine Maintenance and refit the spark plug, tightening it to the specified torque setting.

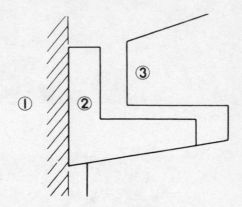

Fig. 1.25 Top piston ring position – TY125 and 175

1 Cylinder   3 Piston
2 Piston ring

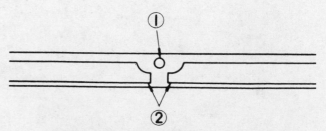

Fig. 1.26 Correct positioning of piston ring end gap

1 Locating peg   2 Piston ring end gaps

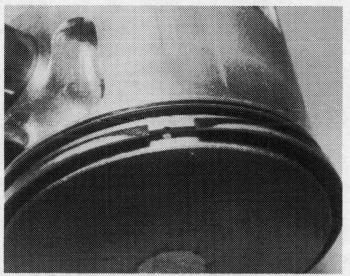

36.2 Dykes top ring is fitted as shown – TY125 and 175 – note ring locating peg

36.3a Lubricate all crankshaft bearings and refit small-end bearing – note rag packing crankcase mouth

36.3b Arrow cast in piston crown faces towards exhaust port

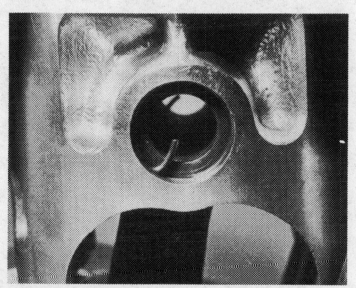

36.4 Ensure circlips are correctly located in grooves – always use new circlips

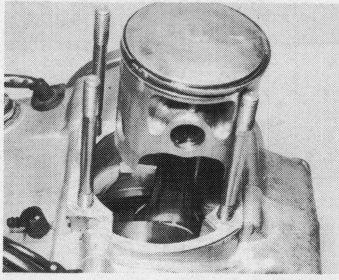

36.5 Fit new base gasket, check rings are correctly located and oil piston before refitting barrel

36.6a Plain copper head gaskets can be reused if annealed – see Section 18

36.6b Tighten cylinder head nuts in diagonal sequence to correct torque setting

36.7 Carburettor stub is slightly offset – refit intake stub as shown

37 Refitting the engine/gearbox unit into the frame

1   Check that no part has been omitted during reassembly. Prepare the machine and ease the engine into position. Align the engine and fit the mounting bolts and nuts; tighten these to the specified torque settings.

2   Route the main generator lead from the crankcase top up to the frame, connect it again to the main loom and secure it with any clamps or ties provided. Connect the HT lead to the spark plug.

3   Engage the sprocket on the chain and fit the sprocket to the output shaft end, followed by a lock washer and the nut. Apply the rear brake hard and tighten the nut to the specified torque setting, then secure the nut by bending up against one of its flats an unused portion of the lock washer.

4   Working as described in Routine Maintenance check, and adjust if necessary, the chain tension, rear brake adjustment and stop lamp rear switch setting.

5   Check, and reset if necessary, the ignition timing as described in Routine Maintenance, then refit the crankcase left-hand cover and the gearchange pedal. On TY50 and TY80 models, adjust the clutch release mechanism as described in Routine Maintenance. On TY125 and TY175 models refit the clutch cable, bending up the metal tang to retain the nipple in the operating lever then adjust the clutch release mechanism.

6   Refit the carburettor to the intake stub, check it is vertical and tighten the clamp or mounting bolts (as appropriate) then refit the air filter hose and tighten its clamp. Insert the throttle slide assembly, ensuring that it fits correctly, then tighten the carburettor top. Refit the oil pump cable to the pump, screwing the adjuster elbow as far as possible into the crankcase before aligning it so that the cable is smoothly routed and tightening the locknut.

7   Working as described in Routine Maintenance, check the throttle cable and the pump cable adjustment. Connect the oil tank/pump feed pipe at the tank union again and bleed any air from the feed pipe and pump using the bleed screw, as described in Chapter 2. Re-route the tank breather pipe, if disturbed, and mop up any spilt oil, then secure the tank/pump feed pipe with the clamps provided. Temporarily refit the oil pump/engine feed pipe to its union on the carburettor or intake stub but remember that it is still to be cleared of air bubbles. Top up the oil tank if necessary.

8   Refit the exhaust system. Use grease to stick a new gasket to the exhaust port and fit the mountings loosely. Tighten the exhaust port fasteners first, then the remaining mountings. Where applicable, do not forget to renew the O-ring sealing the joint between the exhaust front and rear sections. Refit the kickstart lever, tightening its pinch bolt securely.

9   Where applicable, reconnect the battery. Check the terminals are clean and prevent corrosion occurring by smearing them with petroleum jelly. Route the vent pipe clear of the lower frame.

10   Refit the fuel tank, checking there is no metal-to-metal contact which will split the tank. Reconnect the fuel feed pipe, turn on the tap and check for leaks which must be cured immediately. Where fitted, re-route the filler cap breather pipe.

11   Refit the seat, the side panel and the crankcase bashplate (as applicable), check that the transmission oil drain plug is refitted and tightened to the specified torque setting, then refill the gearbox with oil to the correct level as described in Routine Maintenance. Note however, that slightly more oil may be required after a full engine rebuild; refer to the Specification Section of Chapter 2.

12   Make a final check that all components have been refitted and that all are securely fastened and correctly adjusted.

37.3a Do not forgt to refit output shaft spacer ...

37.3b ... before refitting the gearbox sprocket ...

37.3c ... followed by its lock washer

37.3d The sprocket retaining nut must be fitted as shown to clear the shaft splines

37.3e Secure the nut with the lock washer as shown to prevent it from slackening

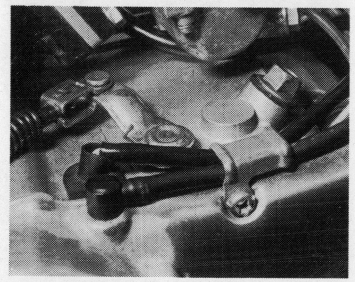

37.5a Route carefully the electrical lead and breather hose as shown when refitting the cover

37.5b Tighten securely the gearchange lever pinch bolt

37.5c Bend down the small tang to secure the clutch cable end nipple – TY125 and 175

37.6 Screw the oil pump cable adjuster elbow as far as possible into the crankcase

37.8 Always fit a new exhaust gasket

37.11 Fill the gearbox with the correct amount of oil and check the level after the engine has been restarted

38 Starting and running the rebuilt engine

1 Start the engine using the usual procedure adopted for a cold engine. Do not be disillusioned if there is no sign of life initially. A certain amount of perseverance may prove necessary to coax the engine into activity even if new parts have not been fitted. Should the engine persist in not starting, check that the spark plug has not become fouled by the oil used during re-assembly. Failing this, go through the fault-finding charts and work out what the problem is methodically.

2 When the engine does start, keep it running as slowly as possible to allow the oil to circulate. Open the choke as soon as the engine will run without it. During the initial running, a certain amount of smoke may be in evidence due to the oil used in the reassembly sequence being burnt away. The resulting smoke should gradually subside. As soon as the engine will run smoothly, carry out the procedure described in Chapter 2 to bleed air from the oil pump/engine feed pipe, then connect the pipe firmly to its union and secure it with the tubular metal clip.

3 Check the engine for blowing gaskets and oil leaks. Before using the machine on the road, check that all the gears select properly, and that the controls function correctly.

38.2a Secure oil feed pipe with tubular clip after pipe has been cleared of air

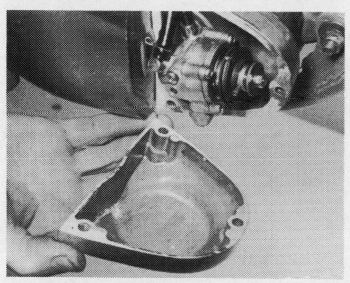

38.2b Fit a new gasket and lubricate pump cable before refitting pump cover

39 Taking the rebuilt machine on the road

1   Any rebuilt machine will need time to settle down, even if parts have been replaced in their original order. For this reason it is highly advisable to treat the machine gently for the first few miles to ensure oil has circulated throughout the lubrication system and that new parts fitted have begun to bed down.

2   Even greater care is necessary if the engine has been rebored or if a new crankshaft has been fitted. In the case of a rebore, the engine will have to be run in again, as if the machine were new. This means greater use of the gearbox and a restraining hand on the throttle until at least 500 miles have been covered. There is no point in keeping to any set speed limit; the main requirement is to keep a light loading on the engine and to gradually work up performance until the 500 mile mark is reached. These recommendations can be lessened to an extent when only a new crankshaft is fitted. Experience is the best guide since it is easy to tell when an engine is running freely.

3   Remember that a good seal between the piston and the cylinder barrel is essential for the correct functioning of the engine. A rebored two-stroke engine will require more careful running-in, over a long period, than its four-stroke counterpart. There is a far greater risk of engine seizure during the first hundred miles if the engine is permitted to work hard.

4   If at any time a lubrication failure is suspected, stop the engine immediately and investigate the cause. If an engine is run without oil, even for a short period, irreparable engine damage is inevitable.

5   Do not on any account add oil to the petrol under the mistaken belief that a little extra oil will improve the engine lubrication. Apart from creating excess smoke, the addition of oil will make the mixture much weaker, with the consequent risk of overheating and engine seizure. The oil pump alone should provide full engine lubrication.

6   Do not tamper with the exhaust system or run the engine without the baffle fitted to the silencer. Unwarranted changes in the exhaust system will have a marked effect on engine performance, invariably for the worse. The same advice applies to dispensing with the air cleaner or the air cleaner element.

7   When the initial run has been completed allow the engine unit to cool and then check all the fittings and fasteners for security. Re-adjust any controls which may have settled down during initial use and check the transmission oil level, topping up if necessary.

Chapter 2 Fuel system and lubrication

Contents

Specifications

Fuel tank capacity

	TY50	TY80	TY125, TY175
Overall	4.7 lit (1.03 gal)	2.5 lit (0.55 gal)	4.0 lit (0.88 gal)

Carburettor

	TY50P, TY50M-early	TY50M-late	TY80	TY125	TY175
Manufacturer	Mikuni	Mikuni	Teikei	Mikuni	Mikuni
Type	VM16SH	VM16SH	Y16P-3	VM22SS	VM22SS
ID number	1G790	2K000	45160	1K600	52561
Choke size	16mm (0.63in)	16mm (0.63in)	16mm (0.63in)	22mm (0.87in)	22mm (0.87in)
Main jet	65	70	86	100	240
Air jet	2.5	1.0	N/Av	2.5	2.5
Pilot jet	20	20	34	25	25
Needle jet	E-4	E-2	2-080	0-0	0-0
Jet needle	3D3	3D3	049	4L6	4L6
Clip position - grooves from top	2nd	2nd	2nd	4th	4th
Throttle slide (valve) cutaway	2.0	2.0	1.0	3.0	3.0
Starter jet	30	30	N/Av	40	N/Av
Float valve seat	1.2	1.2	1.4	1.8	1.8
Float height	22.5±0.5mm (0.89±0.02in)	22.5±0.5mm (0.89±0.02in)	23.0±1.0mm (0.91±0.04in)	21.0±2.5mm (0.83±0.10in)	21.0±2.5mm (0.83±0.10in)
Pilot screw - turns out	$1\frac{1}{4}$	2	$1\frac{1}{2}$	$1\frac{1}{4}$	$1\frac{1}{2}$
Idle speed	1300±50rpm	1300±50rpm	1300±50rpm	1100±50rpm	1100±50rpm

Reed valve

Reed petal thickness	0.15 mm (0.0059 in)
Petal maximum warpage	0.30 mm (0.0118 in)

Stopper plate height:

TY50, 80	7.0 ± 0.2 mm (0.2756 ± 0.0079 in)
TY125	5.0 ± 0.3 mm (0.1969 ± 0.0118 in)
TY175	9.0 ± 0.7 mm (0.3543 ± 0.0276 in)

Engine lubrication system

Type	Pump fed total loss system (Yamaha Autolube)

Oil tank capacity:

TY50	1.20 lit (2.11 pint)
TY80	0.22 lit (0.39 pint)
TY125, TY175	0.30 lit (0.53 pint)
Recommended oil	Good quality self-mixing 2-stroke engine oil

Oil pump colour code:

TY50, TY125	Yellow
TY80, TY175	N/Av

Oil pump minimum stroke:

TY50, TY125, TY175	0.20 - 0.25 mm (0.0079 - 0.0098 in)
TY80	0.30 - 0.35 mm (0.0118 - 0.0138 in)

Oil pump output - per 200 strokes:

	TY50	TY125
At minimum throttle	0.50-0.63cc (0.0176-0.0222 fl oz)	0.95-1.20cc (0.0334-0.0422 fl oz)
At full throttle	4.23-4.27cc (0.1489-0.1503 fl oz)	8.80-9.70cc (0.3098-0.3414 fl oz)

Gearbox lubrication

Recommended oil	SAE 10W/30SE engine oil

Quantity:

TY50, TY80 - at oil change	500 - 550cc (0.88 - 0.97 pint)
TY50, TY80 - at rebuild	550 - 600cc (0.97 - 1.05 pint)
TY125, TY175	650cc (1.14 pint)

Torque wrench settings

Component	kgf m	lbf ft
Transmission oil drain plug:		
TY50, TY80	3.5 - 4.0	25.0 - 29.0
TY125, TY175	2.0 - 2.5	14.5 - 18.0
Reed valve assembly mounting bolts	0.8	6.0
Reed petal mounting screws	0.08	0.6

1 General description

The fuel system comprises a petrol tank, from which petrol is fed by gravity to the float chamber of the carburettor, via a three position petrol tap. The tap has 'Off', 'On' and 'Reserve' positions, the latter providing a warning that the petrol level is low in time for the owner to find a garage.

The carburettor is of conventional concentric design, the float chamber being integral with the lower part of the carburettor. Cold starting is assisted by a separate starting circuit which supplies the correct fuel-rich mixture when the 'choke' control is operated.

Air entering the carburettor passes through an air cleaner casing which contains an oil-impregnated foam air filter. This effectively removes any airborne dust, which would otherwise enter the engine and cause premature wear. The air cleaner also helps silence induction noise, a common problem with two-stroke engines.

A reed valve is fitted in the intake tract to help the engine's performance at low to medium speeds. Consisting of two reed 'petals' mounted on an alloy valve body, the petals open and close automatically as a result of the difference between atmospheric pressure and the pressure in the crankcase as the engine is running. Their movement is controlled by two carefully-shaped stopper plates.

Engine lubrication is catered for by the Yamaha Autolube system. Oil from a separate tank is fed by an oil pump to a small injection nozzle in the carburettor body or the intake tract. The pump is linked to the throttle twistgrip, and this controls the volume of oil fed to the engine.

All gearbox and primary drive components are lubricated by splash from a supply of oil contained in the reservoir formed by the crankcase castings.

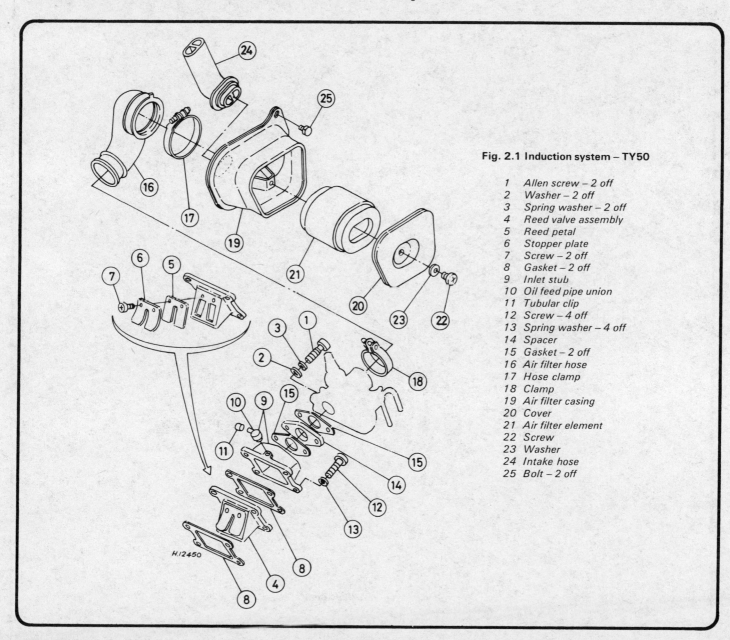

Fig. 2.1 Induction system – TY50

1   Allen screw – 2 off
2   Washer – 2 off
3   Spring washer – 2 off
4   Reed valve assembly
5   Reed petal
6   Stopper plate
7   Screw – 2 off
8   Gasket – 2 off
9   Inlet stub
10 Oil feed pipe union
11 Tubular clip
12 Screw – 4 off
13 Spring washer – 4 off
14 Spacer
15 Gasket – 2 off
16 Air filter hose
17 Hose clamp
18 Clamp
19 Air filter casing
20 Cover
21 Air filter element
22 Screw
23 Washer
24 Intake hose
25 Bolt – 2 off

H.12450

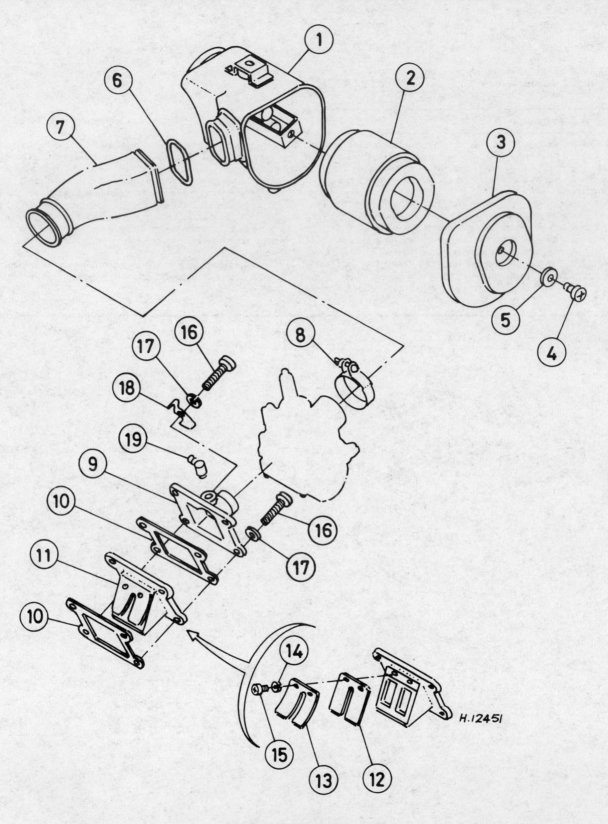

Fig. 2.2 Induction system – TY80

1	Air filter casing	6	Spring clamp	10	Gasket – 2 off
2	Air filter element	7	Air filter hose	11	Reed valve assembly
3	Cover	8	Hose clamp	12	Reed petal
4	Screw	9	Inlet stub	13	Stopper plate
5	Washer				

14   Spring washer – 2 off
15   Screw – 2 off
16   Screw – 4 off
17   Spring washer – 4 off

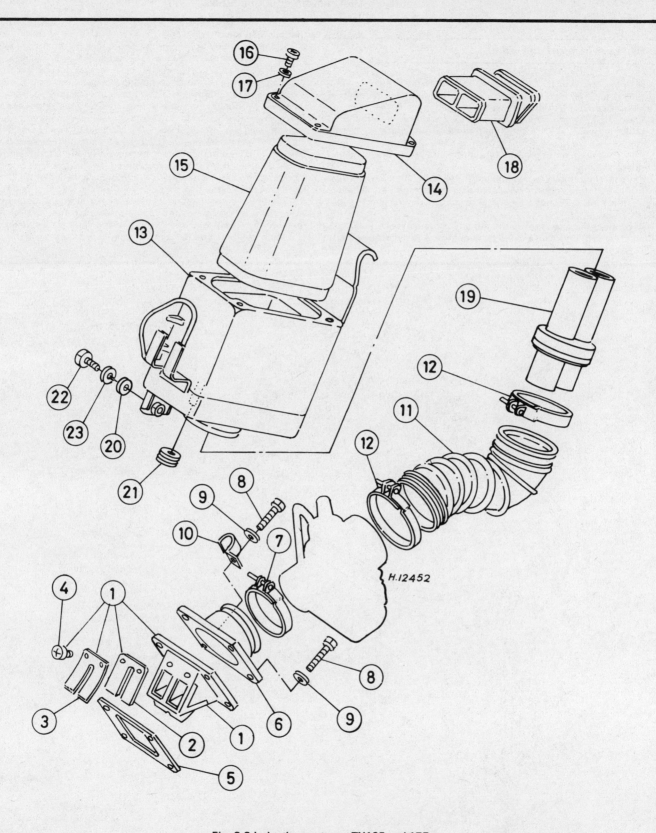

Fig. 2.3 Induction system – TY125 and 175

1	Reed valve assembly	7	Clamp	13	Air filter casing	19	Baffle tube
2	Reed petal	8	Bolt – 4 off	14	Cover	20	Washer
3	Stopper plate – 2 off	9	Washer – 4 off	15	Air filter element	21	Grommet
4	Screw – 4 off	10	Hose clip	16	Screw – 4 off	22	Bolt
5	Gasket	11	Air filter hose	17	Washer – 4 off	23	Spring washer
6	Inlet stub	12	Hose clamp – 2 off	18	Intake duct		

2 Fuel tank: removal and refitting

1   Turn the tap lever to 'Off'. Observe the necessary fire precautions and unclip the fuel feed pipe, draining any fuel from the pipe into a clean container. Where fitted, release the filler cap breather pipe.

2   Raise the seat and remove the tank retaining bolt or rubber. Pull the tank rearwards off its mounting rubbers and place it in safe storage, away from naked flame or sparks. Check for leaks around the tap and cover the tank to protect its paint finish. Renew any damaged mounting rubbers.

3   Reverse the removal procedure to fit the tank. If necessary, wipe the front mounting rubbers with petrol to ease their reinstallation. Secure the tank and check for metal-to-metal contact which might split the tank.

4   Reconnect the pipe, turn on the tap and check for leaks. Fuel leaks will waste petrol and cause a fire hazard. Where applicable, re-route the filler cap breather pipe.

5   If it is necessary to remove the petrol tank for repairs the following points should be noted. Petrol tank repair, whether necessitated by accident damage or by petrol leaks, is a task for the professional. Welding or brazing is not recommended unless the tank is purged of all petrol vapour; which is a difficult condition to achieve. Resin-based tank sealing compounds are a much more satisfactory method of curing leaks, and are now available through suppliers who advertise regularly in the motorcycle press. Accident damage repairs will inevitably involve re-painting the tank; matching of modern paint finishes is a very difficult task not to be lightly undertaken by the average owner. It is therefore recommended that the tank be removed by the owner, and then taken to a motorcycle dealer or similar expert for professional attention.

6   Repeated contamination of the fuel tap filter and carburettor by water or rust and paint flakes indicates that the tank should be removed for flushing with clean fuel and internal inspection. Rust problems can be cured by using a resin-based tank sealant.

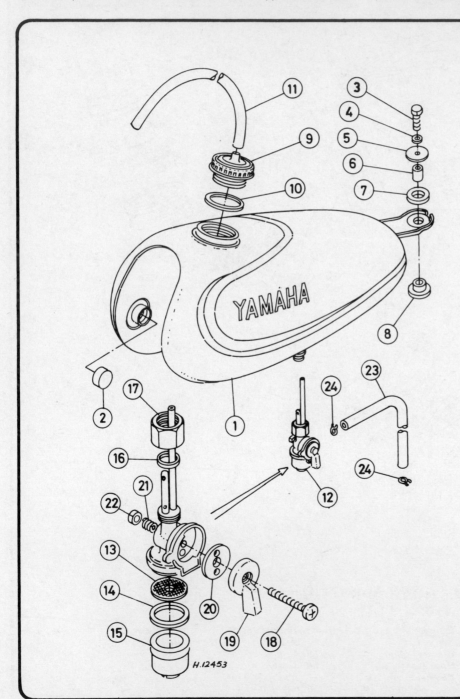

Fig. 2.4 Fuel tank and tap – TY50

1   Fuel tank
2   Mounting rubber
3   Bolt
4   Spring washer
5   Washer
6   Spacer
7   Mounting rubber
8   Mounting rubber
9   Filler cap
10 Gasket
11 Breather pipe
12 Fuel tap assembly
13 Filter gauze
14 Sealing washer
15 Filter bowl
16 Seal
17 Gland nut
18 Screw
19 Tap lever
20 Seal
21 Spring
22 Nut
23 Fuel pipe
24 Clip – 2 off

H.12453

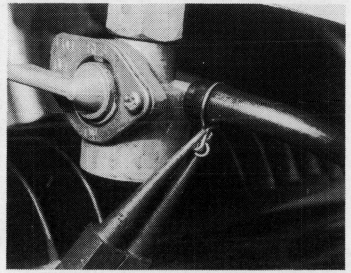

2.1 Use pliers to release wire retaining clip before disconnecting fuel feed pipe

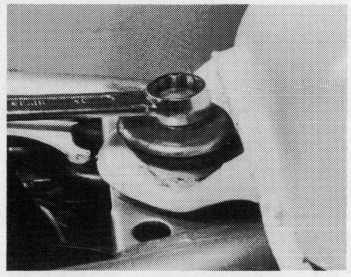

2.2 Raise or remove seat to expose tank rear mounting bolt

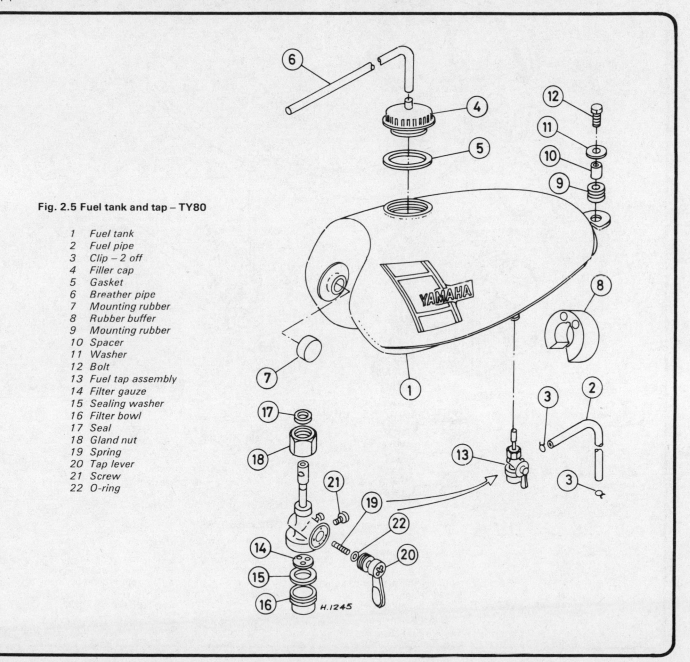

Fig. 2.5 Fuel tank and tap – TY80

1 Fuel tank
2 Fuel pipe
3 Clip – 2 off
4 Filler cap
5 Gasket
6 Breather pipe
7 Mounting rubber
8 Rubber buffer
9 Mounting rubber
10 Spacer
11 Washer
12 Bolt
13 Fuel tap assembly
14 Filter gauze
15 Sealing washer
16 Filter bowl
17 Seal
18 Gland nut
19 Spring
20 Tap lever
21 Screw
22 O-ring

H.1245

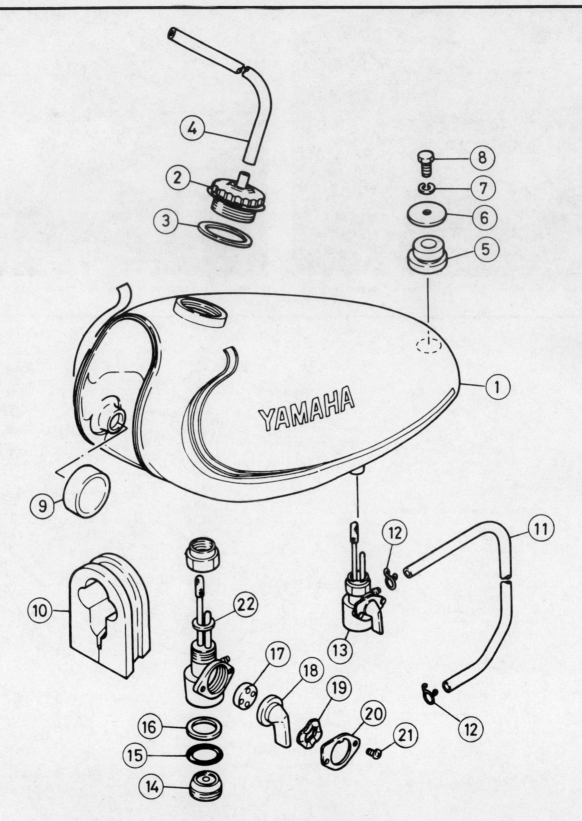

Fig. 2.6 Fuel tank and tap – TY125 and 175

1	Fuel tank	7	Spring washer	13	Fuel tap assembly	18	Tap lever
2	Cap	8	Bolt	14	Filter bowl	19	Wave washer
3	Gasket	9	Rubber mounting	15	O-ring	20	Lever retaining plate
4	Hose	10	Rubber mounting	16	Sealing washer	21	Screw
5	Rubber mounting	11	Fuel pipe	17	Seal	22	Gasket
6	Washer	12	Clip				

3 Fuel tap: removal, dismantling and reassembly

1 The tap is secured to the tank by a large gland nut. To remove it, drain the tank, removing the latter from the machine if required, then unscrew the nut and withdraw the tap. Unscrew the filter bowl from the bottom of the tap body and remove the sealing O-ring and washer.

2 Remove the single retaining screw (TY50, TY80) or the two retaining screws (TY125, TY175) and withdraw the tap lever with the retaining plate and wave washer, then carefully prise out the tap seal.

3 If any leaks have occurred, renew the seal concerned; in the case of the lever seal a temporary repair can be effected sometimes by reversing the seal.

4 Clean out the tap passages using compressed air and clean the gauze filters with a fine toothbrush, taking great care not to damage the gauze.

5 On reassembly, take care not to overtighten any component; the castings are of a zinc-based alloy which fractures easily. Never overtighten a component in an effort to cure a fuel leak.

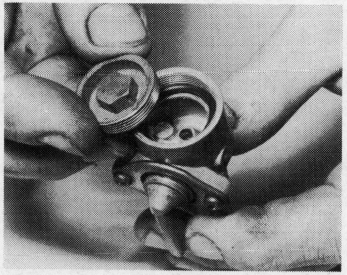

3.1a Unscrew filter bowl from base of tap body ...

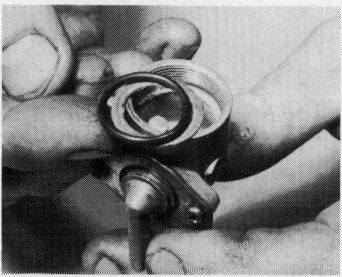

3.1b ... and note O-ring and sealing washer – renew if damaged or worn

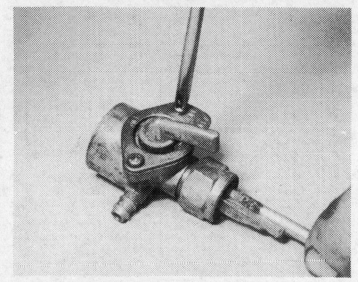

3.2a Remove two retaining screws to release retaining plate ...

3.2b ... with the tap lever and wave washer – TY125 and 175

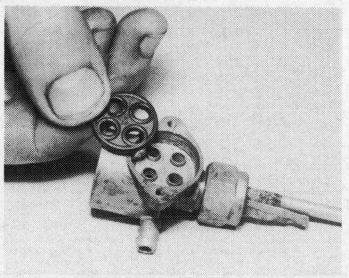

3.2c Withdraw tap seal – renew if damaged or worn

4 Fuel feed pipe: examination

1   The fuel feed pipe is made from thin walled synthetic rubber and is of the push-on type. It is necessary to replace the pipe only if it becomes hard or splits. It is unlikely that the retaining clips will need replacing due to fatigue as the main seal between the pipe and union is an interference fit.

2   If the pipe has been replaced with the transparent plastic type for any reason, look for signs of yellowing which indicate that the pipe is becoming brittle due to the plasticiser being leached out by the petrol. It is a sound precaution to renew a pipe when this occurs, as any subsequent breakage whilst in use will be almost impossible to repair. **Note**: On no account should natural rubber tubing be used to carry petrol, even as a temporary measure. The petrol will dissolve the inner wall, causing blockages in the carburettor jets which will prove very difficult to remove.

5 Carburettor: removal and refitting

1   As a general rule, the carburettor should be left alone unless it is in obvious need of overhaul. Before a decision is made to remove and dismantle, ensure that all other possible sources of trouble have been eliminated. This includes the more obvious candidates such as a fouled spark plug, a dirty air filter element or choked exhaust system. If a fault has been traced back to the carburettor, proceed as follows.

2   Make sure that the fuel tap is turned off, then prise off the feed pipe at the carburettor union. Where applicable the oil feed pipe is removed in a similar manner, noting that the small tubular clip should be displaced first. The pipe can then be eased away from its union with the aid of an electrical screwdriver.

3   Slacken the screws of the clips which secure the carburettor to its inlet stub and air filter hose so that the carburettor can be twisted free of them and partially removed. On TY50 and TY80 models remove the two mounting bolts then follow the same procedure. This affords access to the threaded carburettor top, which should be unscrewed to allow the throttle valve (slide) assembly to be withdrawn. It is not normally necessary to remove this from the cable and it can be left attached and taped clear of the engine. If removal is necessary, however, proceed as follows.

4   Holding the carburettor top, compress the throttle return spring against it and hold it in position against the cap; the cable can be pushed down and slid out of its locating groove. The various parts can now be removed and should be placed with the carburettor.

5   The carburettor is refitted by reversing the removal sequence. Note that it is important that the instrument is mounted vertically to ensure that the fuel level in the float bowl is correct. On refitting the throttle valve assembly, ensure that the cutaway faces towards the air filter and that the groove in the slide is engaged correctly with the projection in the carburettor body. Once refitted, check the carburettor adjustments as described later in this Chapter. Note too that the oil pump delivery pipe should be bled and the pump adjustments checked after overhaul.

6   **Note**: If the carburettor is to be set up from scratch it is important to check jet and float level settings prior to installation. To this end, refer to Sections 7 and 8 before the carburettor is refitted.

6 Carburettor: dismantling, examination and reassembly

1   Cover an area of work surface with clean paper. This will prevent components placed upon it from becoming dirty or lost.

2   On TY175 models only, unscrew the main jet holder from the bottom of the float chamber and unscrew the main jet. On all models, remove the float chamber retaining screws. If necessary, tap around the chamber to body joint with a soft-faced hammer to free the chamber. Remove the float pivot pin and detach the twin floats. Displace the float needle and unscrew the float needle seat with sealing washer.

3   Unscrew the pilot jet, main jet and needle jet, using only a close fitting spanner or screwdriver otherwise damage will occur to the soft brass jets.

4   Note the pilot and throttle stop screw settings by counting the number of turns required to screw them fully in until seating lightly; this will make it easier to 'retune' the carburettor after reassembly. Remove both screws with their springs.

5   Refer to the relevant figure accompanying this text and remove the choke assembly. It is not necessary to remove the drain plug from the float chamber except for seal renewal.

6   Before examination, thoroughly clean each part in clean petrol, using a soft nylon brush to remove stubborn contamination and a compressed air jet to blow dry. Avoid using rag because lint will obstruct jet orifices. Do not use wire to clear blocked jets, this will enlarge the jet and increase petrol consumption; if an air jet fails use a soft nylon bristle. Observe the necessary fire precautions and wear eye protection against blow-back from the air jet.

7   Check carefully for any distorted or cracked castings, renew all O-rings, gaskets, fatigued or broken springs and flattened spring washers.

8   Wear of the float needle takes the form of a groove around its seating area; renew if worn. Check for similar wear of the needle seat and, if possible, renew when worn. Where fitted, check the needle end pin is free to move and is spring loaded.

9   Check the floats for damage and leakage. Renew if damaged, it is not advisable to attempt a repair.

10 Examine the choke assembly, renewing any worn or damaged parts. Renew hardened drain or fuel feed pipes.

11 Wear of the throttle valve will be indicated by polished areas on its external diameter, causing air leaks which weaken the mixture and produce erratic slow running. Examine the carburettor body for similar wear and renew each component as necessary.

12 Examine the needle for scratches or wear along its length and for straightness. If necessary, it must be renewed, in conjunction with the needle jet. Do not attempt to straighten a bent needle; they fracture very easily.

13 Renew the seal within the mixing chamber top if damaged. The throttle return spring must be free of fatigue or corrosion.

14 Before assembly, clean all parts and place them on clean paper in a logical order. Do not use excessive force during reassembly, it is easy to shear a jet or damage a casting. When refitting the float bowl on TY175 models apply a smear of oil to the O-ring at the base of the needle jet to help it fit into the recess in the float bowl.

15 Reassembly is the reverse of dismantling. If in doubt, refer to the accompanying figures or photographs. Seat the pilot and throttle stop screws lightly before screwing out to their previously noted settings.

5.3 Slacken (or remove) carburettor mountings to withdraw carburettor body

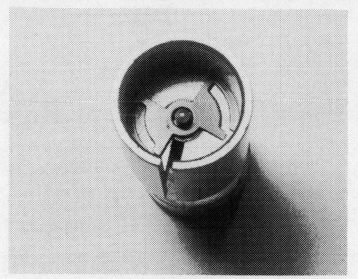

5.4a Throttle needle is located in valve (slide) by retainer plate

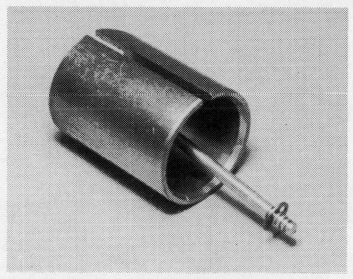

5.4b Check position of needle clip in needle grooves

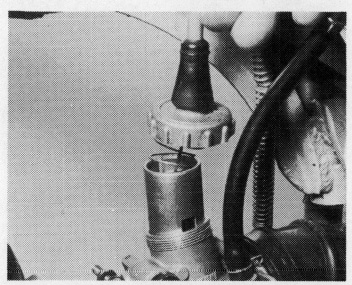

5.5 Ensure throttle valve (slide) is correct way round on refitting

6.2a TY175 — main jet is mounted in separate holder in base of float chamber ...

6.2b ... remove jet and carefully clean holder passages

6.2c Remove screws to release float chamber ...

6.2d ... and withdraw pivot pin to release float assembly ...

6.2e ... followed by the float needle – check tip for signs of wear or damage

6.2f Remove float valve seat – renew sealing washer if excessively flattened or worn

6.3 Use only close fitting spanner or screwdriver to remove jets – pilot jet shown

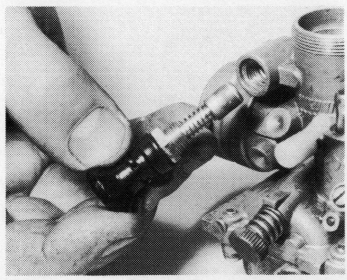

6.5 Choke plunger is removed as a single unit – can be dismantled if necessary

6.14 TY175 only -- apply a smear of oil to O-ring at base of needle jet when refitting float chamber

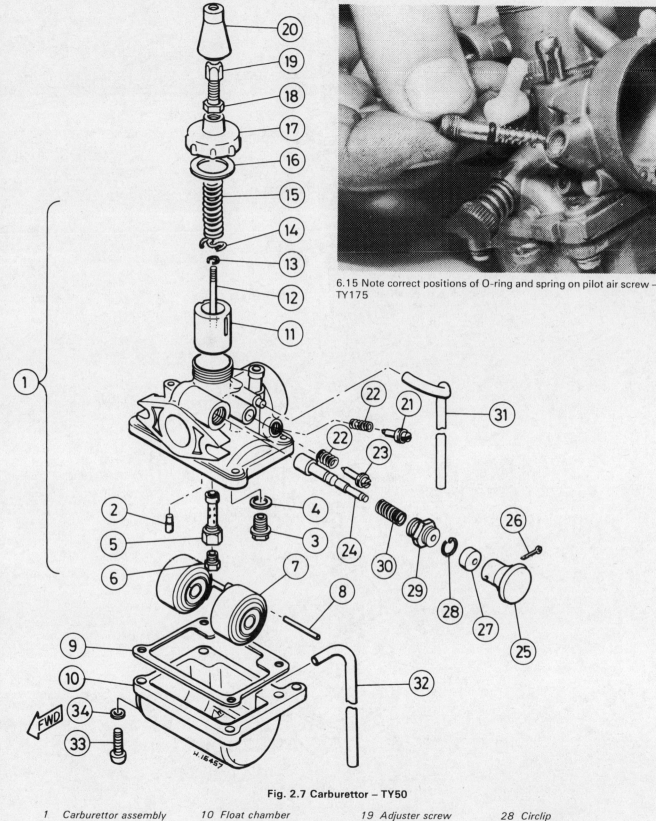

6.15 Note correct positions of O-ring and spring on pilot air screw – TY175

Fig. 2.7 Carburettor – TY50

1	Carburettor assembly	10	Float chamber	19	Adjuster screw	28	Circlip
2	Pilot jet	11	Throttle valve	20	Rubber sleeve	29	Mounting nut
3	Float needle and seat	12	Jet needle	21	Pilot air screw	30	Spring
4	Sealing washer	13	Needle clip	22	Spring – 2 off	31	Breather hose
5	Needle jet	14	Needle retaining plate	23	Throttle stop screw	32	Breather hose
6	Main jet	15	Spring	24	Choke plunger	33	Screw – 4 off
7	Float	16	Sealing washer	25	Knob	34	Spring washer 4 off
8	Float pivot pin	17	Carburettor top	26	Split pin		
9	Gasket	18	Locknut	27	Rubber cap		

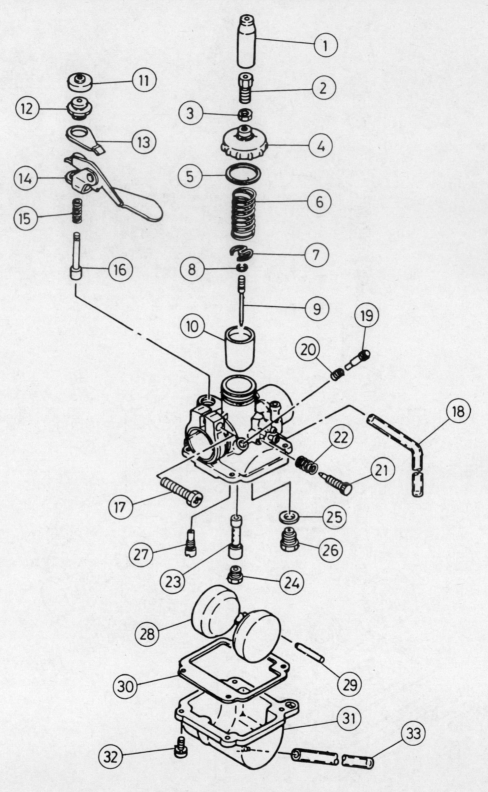

Fig. 2.8 Carburettor – TY80

1 Rubber sleeve	10 Throttle valve	19 Pilot mixture screw	28 Float
2 Adjuster screw	11 Rubber cap	20 Spring	29 Float pivot pin
3 Locknut	12 Knob	21 Throttle stop screw	30 Gasket
4 Carburettor top	13 Spring plate	22 Spring	31 Float chamber
5 Sealing washer	14 Lever assembly	23 Needle jet	32 Screw – 4 off
6 Spring	15 Spring	24 Main jet	33 Breather hose
7 Needle retainer	16 Plunger	25 Sealing washer	
8 Needle clip	17 Clamp bolt	26 Float needle and seat	
9 Jet needle	18 Breather hose	27 Pilot jet	

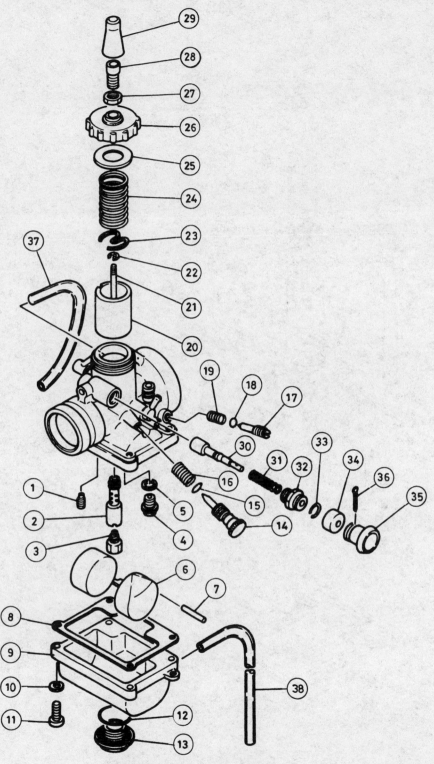

Fig. 2.9 Carburettor – TY125

1 Pilot jet	11 Screw	21 Jet needle	31 Spring
2 Needle jet	12 O-ring	22 Clip	32 Plunger cap
3 Main jet	13 Drain plug	23 Needle retaining plate	33 Clip
4 Float needle and seat	14 Throttle stop screw	24 Throttle valve spring	34 Plunger cap cover
5 Sealing washer	15 O-ring	25 Seal	35 Knob
6 Float	16 Spring	26 Carburettor top	36 Split pin
7 Pivot pin	17 Pilot air screw	27 Locknut	37 Pipe
8 Gasket	18 O-ring	28 Adjuster	38 Pipe
9 Float chamber	19 Spring	29 Rubber sleeve	
10 Spring washer	20 Throttle valve	30 Choke plunger	

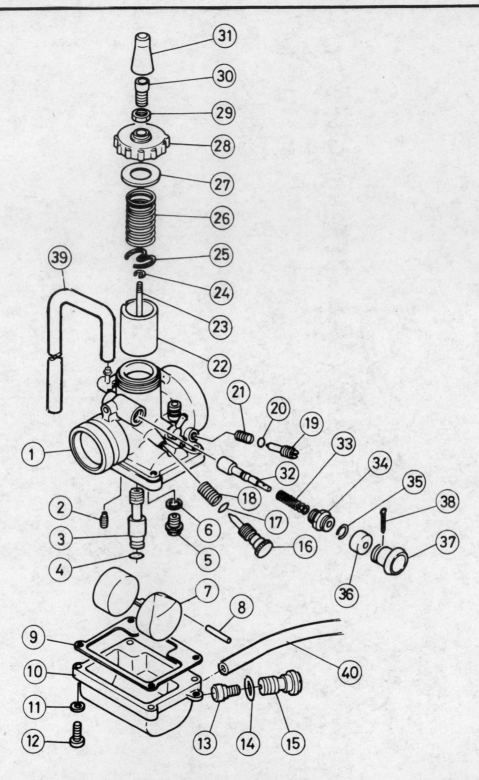

Fig. 2.10 Carburettor – TY175

1 Carburettor	11 Spring washer – 4 off	21 Spring	31 Rubber sleeve
2 Pilot jet	12 Screw – 4 off	22 Throttle valve	32 Choke plunger
3 Needle jet	13 Main jet	23 Jet needle	33 Spring
4 O-ring	14 O-ring	24 Needle clip	34 Mounting nut
5 Float needle and seat	15 Main jet holder	25 Needle retaining plate	35 Circlip
6 Sealing washer	16 Throttle stop screw	26 Spring	36 Rubber cap
7 Float	17 O-ring	27 Seal	37 Knob
8 Float pivot pin	18 Spring	28 Carburettor top	38 Split pin
9 Gasket	19 Pilot air screw	29 Locknut	39 Breather hose
10 Float chamber	20 O-ring	30 Adjuster screw	40 Breather hose

7 Carburettor: checking the settings

1 The various jet sizes, throttle valve cutaway and needle position are predetermined by the manufacturer and should not require modification. Check with the Specifications list at the beginning of this Chapter if there is any doubt about the types fitted. If a change appears necessary it can often be attributed to a developing engine fault unconnected with the carburettor. Although carburettors wear in service, this process occurs slowly over an extended length of time and hence wear of the carburettor is unlikely to cause sudden or extreme malfunction. If a fault does occur check first other main systems, in which a fault may give similar symptoms, before proceeding with carburettor examination or modification.

2 Where non-standard items, such as exhaust systems and air filters have been fitted to a machine, some alterations to carburation may be required. Arriving at the correct settings often requires trial and error, a method which demands skill born of previous experience. In many cases the manufacturer of the non-standard equipment will be able to advise on correct carburation changes.

3 As a rough guide, up to $\frac{1}{8}$ throttle is controlled by the pilot jet, $\frac{1}{8}$ to $\frac{1}{4}$ by the throttle valve cutaway, $\frac{1}{4}$ to $\frac{3}{4}$ throttle by the needle position and from $\frac{3}{4}$ to full by the size of the main jet. These are only approximate divisions, which are by no means clear cut. There is a certain amount of overlap between the various stages.

4 If alterations to the carburation must be made, always err on the side of a slightly rich mixture. A weak mixture will cause the engine to overheat, particularly on two-stroke engines, may cause engine seizure. Reference to Routine Maintenance will show how, after some experience has been gained, the condition of the spark plug electrodes can be interpreted as a reliable guide to mixture strength.

8 Carburettor: adjustment

1 Before any dismantling or adjustment is undertaken, eliminate all other possible causes of running problems, checking in particular the spark plug, ignition timing, air cleaner and the exhaust. Checking and cleaning these items as appropriate will often resolve a mysterious flat spot or misfire.

2 The first step in carburettor adjustment is to ensure that the jet sizes, needle position and float height are correct, which will require the removal and dismantling of the carburettor as described in Sections 5 and 6 of this Chapter.

3 If the carburettor has been removed for the purpose of checking jet sizes, the float level should be measured at the same time. It is unlikely that once this is set up correctly there will be a significant amount of variation, unless the float needle or seat have worn. These should be checked and renewed, if necessary, as described in Section 6.

4 Remove the float bowl from the carburettor body, if this has not already been done, and very carefully peel away the float chamber gasket. Check that the gasket surface of the carburettor body is clean and smooth once the gasket is removed. Hold the carburettor body so that the venturi is now vertical with the air filter side upwards and the floats are hanging from their pivot pin. Carefully tilt the carburettor to an angle of about 60 – 70° from the vertical so that the tang of the float pivot is resting firmly on the float needle and the float valve is therefore closed, but also so that the spring-loaded pin set in the float needle itself is not compressed. Measure the distance between the gasket face and the bottom of one float with an accurate ruler or a vernier caliper; the distance should be as specified for the machine being worked on. A tolerance above or below the set figure is allowed, but the more accurate the setting, the better the engine's performance, reliability and economy will be.

5 If adjustment is required, remove the float assembly and bend by a very small amount the small tang which acts on the float needle pin. Reassemble the float and measure the height again. Repeat the process until the measurement is correct, then check that the other float is exactly the same height as the first. Bend the pivot very carefully and gently if any difference is found between the heights of the two floats.

6 When the jet sizes have been checked and reset as necessary, reassemble the carburettor and refit it to the machine as described in Sections 5 and 6 of this Chapter.

7 Start the engine and allow it to warm up to normal operating temperature, preferably by taking the machine on a short journey. Stop the engine and screw the pilot screw in until it seats lightly, then unscrew it by the number of turns shown in the Specifications Section for the particular model. Start the engine and set the machine to its specified idle speed by rotating the throttle stop screw as necessary. Note that the idle speed should be regarded as the slowest speed at which the engine will tick over smoothly and reliably. Try turning the pilot screw inwards by about $\frac{1}{4}$ turn at a time, noting its effect on the idling speed, then repeat the process, this time turning the screw outwards.

8 The pilot screw should be set in the position which gives the fastest consistent tickover. The tickover speed may be reduced further, if necessary, by unscrewing the throttle stop screw the required amount. Check that the engine does not falter and stop after the throttle twistgrip has been opened and closed a few times.

9 The advice in the previous two paragraphs applies principally to road machines but the basic principles are the same for trials machines. The main difference is that far more care is required; trials machines must pull smoothly from near-zero engine speeds and also perform well when flat out. The ability to set up a carburettor to achieve this range of performance is only acquired after a great deal of practical experience; if in any doubt take the machine to a trials expert and watch his methods closely. Note however that most riders have individual preferences; while some experts prefer to have the pilot screw setting a little on the rich side so that the engine does not idle but dies after one or two revolutions when the throttle is closed, others prefer the settings on the weak side so that the engine only picks up smoothly when fully warmed up.

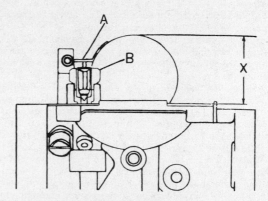

Fig. 2.11 Checking the float height

A   Adjustment tang   X   Float height
B   Float needle

8.7a Adjusting pilot mixture setting using air screw – TY175

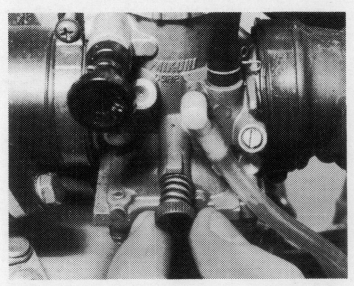

8.7b Use throttle stop screw to set idle speed

9 Reed valves: removal, examination and renovation

1   The reed valve assembly is a precision component, and as such should not be dismantled unnecessarily. The valves are located in the inlet tract, covered by the carburettor flange.

2   Remove the carburettor as described in Section 5 of this Chapter thus exposing the four bolts retaining the inlet stub and the reed valve assembly to the cylinder. After removing these bolts, the assembly can be carefully lifted away.

3   The valves can now be washed in clean petrol to facilitate further examination. They should be handled with great care, and on no account dropped. The stainless steel reeds should be inspected for signs of cracking or fatigue, and if suspect, should be renewed. Remember that any part of the assembly which breaks off in service will almost certainly be drawn into the engine, causing extensive damage. Make a quick check of the state of the assembly by putting the carburettor side to the lips and sucking hard. The reed petals should seal effectively against their seats, providing considerable resistance to the suction applied, although the manufacturer admits some slight leakage is inevitable and acceptable. The reeds should rest flush against their seats but a slight gap is quite normal.

4   Complete the visual inspection by checking the inlet stub for signs of cracking, perishing, or other deterioration, renewing it if necessary. The distance between the inner edge of the valve stopper and the top edge of the valve case is important as it controls the movement of the reed. If smaller than specified, performance will be impaired. More seriously, if larger than specified, the reed may fracture. If the measurement obtained is outside the tolerances specified, the stopper plate should be renewed.

5   To dismantle the reed valve assembly, first degrease it thoroughly and have a spirit-based felt marker pen ready. Mark each face of the valve case in such a way that they can be identified (eg face 'A' and face 'B'). Working on one face at a time, slacken and remove the two screws retaining the reed stopper plate and the reed and withdraw the two components, marking the outer surface of each one with the identifying mark for that face, so that each component of the reed valve assembly can be refitted in exactly the same position. Note that the manufacturer has assisted this by providing a small cutout in the lower right-hand cover of each stopper plate. Put the components to one side, but store them separately to minimise further the risk of incorrect refitting.

6   Check the condition of the various components. The small screws must be in excellent condition and should be renewed if there is any doubt at all about their condition. The stopper plates should be renewed, as previously mentioned, if the stopper plate/valve case heights are above or below the specified tolerances. Never attempt to bend the stopper plate to alter this distance. While this is accepted practice in certain tuning circles to gain more performance, it is not recommended by the manufacturer as the curve of the stopper plates is carefully designed to give maximum performance consistent with maximum reed life. Any attempt to alter the stopper plate heights by bending the plates will alter this curve and may overstress the reeds to the point where fatigue cracks develop and the reeds break off, falling into the engine with disastrous consequences. If there is any evidence of the stopper plates' having been bent, perhaps by a previous owner, they should be renewed.

7   Carefully examine the valve case itself, looking for cracks, distorted mating surfaces, or excessive wear or damage to the neoprene seating face. Remedial action will depend on the nature of the damage found; some resurfacing work may be possible, using a sheet of fine emery paper placed on a flat surface such as a sheet of plate glass, but due to the design and construction of the valve case, such work will be limited in its effect. Any serious damage or wear, or any damage to the neoprene reed seats will mean that the valve case must be renewed, both in the interests of safety and reliability, and in the interest of maximum performance.

8   The reeds themselves, once removed from the valve case, must be checked very carefully, with a magnifying glass if necessary, for signs of fatigue cracks. If undamaged, they must be measured to ensure that they are not worn or distorted. Each reed should be 0.15 mm (0.006 in) thick; there should be no measurable reduction in this thickness at the reed tips. Lay each reed on a completely flat surface such as a sheet of plate glass with its marked, or outer surface facing upwards and hold it by pressing down at the retaining screw holes. In this position any distortion which has taken place will be revealed immediately. The reed should not be distorted at any point more than 0.3 mm (0.012 in), a measurement which can be checked using feeler gauges. Any signs of damage, wear, or distortion revealed by the above tests will mean that the reeds must be renewed. Repairs are not possible and should never be attempted. Do not attempt to bend the reeds straight, if they are found to be distorted, in case fatigue failure is induced.

9   Reassemble the reed valve assembly using the marks made on dismantling and the cutout provided by the manufacturer to ensure that the reeds and stopper plates are refitted in their original positions. A thread locking compound, such as Loctite, must be applied to the four cross-headed screws, which should be tightened progressively to avoid warping the reed or stopper. Do not omit the locking compound, as the screws retain a component which vibrates many times each second and consequently are prone to loosening if assembled incorrectly. Note that a torque setting of 0.08 kgf m (0.6 lbf ft) is specified for these screws; this should be adhered to if the necessary equipment is available. If not, ensure that all four are tightened securely.

10 Refit the reed valve assembly to the cylinder barrel, using a new gasket at the valve case/cylinder barrel joint and applying a thin coat of jointing compound to seal the valve case/inlet stub joint. Tighten the four mounting bolts progressively and in a diagonal sequence to a torque setting of 0.8 kgf m (6 lbf ft). Refit the carburettor as described in Section 5.

9.1 Reed valve assembly is located in inlet tract

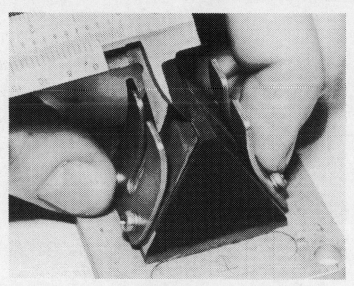

9.4 Measuring reed valve stopper plate height

10 Oil pump: removal, examination and refitting

1 It is rarely necessary to remove the oil pump unless specific attention to it is required. The pump should be considered as a sealed unit – parts are not available and thus it is not practicable to repair it. The pump itself can be removed quite easily leaving the driveshaft and pinion in place in the crankcase right-hand cover. If these latter components require attention it will be necessary to remove the crankcase right-hand cover as described in Section 7 of Chapter 1.

2 The oil pump itself is a very reliable unit which does not give trouble as a rule. If a fault is suspected in the lubrication system, check first the following points:

(a) Sufficient oil in the oil tank, oil tank breather pipes clear
(b) Oil pump correctly adjusted
(c) Oil tank/oil pump and oil pump/carburettor feed pipes free from dirt, kinks, or other obstructions which might be causing a blockage and correctly secured so that air is not entering the system
(d) The check valve ball and spring correctly fitted and functioning properly

3 While the first three of the above items are easy to check and rectify, the last will require some care. The check valve consists of a spring-loaded steel ball which is fitted behind the union of the oil pump/carburettor feed pipe. The valve assembly is secured by the union itself which is a tight press-fit in the pump body and can be removed, therefore, with the careful use of a suitable pair of pliers. Pull the union out, taking care not to damage it, and withdraw the check valve spring and ball. Remove all oil from the components and any particles of dirt, then examine the components, renewing any that are damaged or worn. If necessary, check their condition by comparing them with new components at a Yamaha dealer. Check that the oil pump passage is free from dirt and refit the check valve, taking great care to ensure that the union is a secure, tight fit in the pump body.

4 If the above items have been checked and the oil feed is still suspect, the pump must be removed as follows. First remove the oil pump cover. Disengage the clips of the oil feed pipes and pull each pipe carefully off its union, plugging it immediately with a screw or bolt of suitable size.

5 Disconnect the oil pump cable inner wire from the pump pulley, using, where necessary, a small screwdriver to push back the extension of the pulley return spring or withdrawing the spring clip which retains the inner wire in the pulley track and then disengaging the cable end nipple from the pulley

6 The pump is secured to the cover by two screws. Once these have

been removed the pump can be withdrawn, noting that it may prove necessary to turn the pump slightly to free it from its driveshaft. Note also if a thrust washer is fitted over the end of the driveshaft. This must be picked out of the pump body and replaced on the shaft end so that it is not lost.

7 Further dismantling is not practicable as it will be necessary to renew the pump if it is obviously damaged.

8 Refit the oil pump to the crankcase cover, using a new gasket at the oil pump/crankcase cover joint. Replace and tighten securely the two mounting screws. The remainder of the reassembly is accomplished by reversing the dismantling procedure but note that it will be necessary to check and to reset, if necessary, the pump stroke setting and cable adjustment, and to bleed any air from the pump, as described in Routine Maintenance or Section 11 of this Chapter, respectively.

9 Should it be necessary at any time to inspect the oil pump drive components, the crankcase right-hand cover must be removed as described in Section 7 of Chapter 1. Remove the oil pump and invert the cover. Using a small-bladed screwdriver, prise the drive gear retaining circlip out of its groove in the pump drive shaft and lift the white nylon drive gear away, so that the gear locating pin and the second circlip (where fitted) can be withdrawn. Note the positions of the thrust washers on TY50 and TY80 models. Push the shaft out of its bush in the cover.

10 As the oil pump drive components are lightly loaded and well lubricated, wear is unlikely to occur until a very high mileage has been covered. Damage of any sort should be visible easily and must be rectified by the renewal of the parts concerned. The shaft oil seal may be prised from its housing with a suitably shaped screwdriver. If necessary, the driveshaft bush may be driven out using a hammer and a drift of suitable size, after the casing has been heated by immersing it in boiling water. The state of wear of the driveshaft and bush can be assessed by feeling for excessive free play when the shaft is installed in the bush. Renew the shaft or the bush as necessary.

11 When removing or refitting the bush, and also the oil seal, ensure that the casing is well supported on wooden blocks or a clean, flat surface as appropriate. Complete the assembly procedure by following the reverse of the removal procedure.

11 Oil pump: bleeding

1 It is necessary to bleed the oil pump every time the main feed pipe from the oil tank is removed and refitted. This is because air will be trapped in the oil line, no matter what care is taken when the pipe is removed.

2 Check that the oil pipe is connected correctly, with the retaining clip in position. Then remove the cross-head screw in the outer face of the pump body with the fibre washer beneath the head. This is the oil bleed screw.

3 Check that the oil tank is topped up to the correct level, then place a container below the oil bleed hole to collect the oil that is expelled as the pump is bled. Allow the oil to trickle out of the bleed hole, checking for air bubbles. The bubbles should eventually disappear as the air is displaced by fresh oil. When clear of air, refit the bleed screw.

4 Note also that it will be necessary to ensure that the oil pump/engine feed pipe is primed if this was disturbed. Unless this is checked, the engine will be starved of oil until the pipe fills. The procedure required to avoid this is to start the engine and allow it to idle (maximum speed of 2000 rpm) for a few minutes whilst holding the pump pulley in its fully open position by pulling the pump cable. The excess oil will make the exhaust smoke heavily for a while.

5 Note that earlier models are fitted with a white nylon starter pinion at the rear of the pump unit. This is rotated clockwise, viewed from the rear, to drive the pump independently of the engine for the purpose of setting the pump minimum stroke position and for bleeding air from the pump/engine feed pipe. It was not fitted to later models as it was found that its use was time consuming and unnecessary. Owners of such machines may use either method, as desired; that described above is much quicker.

10.1 Pump assembly can be removed leaving driveshaft in place

10.9a Remove retaining circlip to release oil pump drive components ...

10.9b ... and withdraw pump drive gear ...

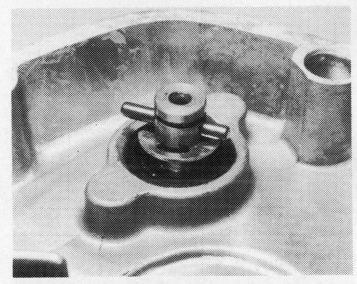

10.9c ... followed by locating pin and remaining circlip or thrust washer

10.9d Pump driveshaft can then be lifted out of the crankcase cover

11.2 Remove oil pump bleed screw ...

11.3 ... and allow oil to trickle out until no more air bubbles can be seen

12 Oil pump: removal for trials use

1 The single most popular modification to the TY Yamahas is to remove the Autolube system, the claimed advantages being as follows:

  a) Loss of weight
  b) Lighter throttle
  c) Easier to set up carburation, especially at low speed
  d) Simplified maintenance

2 Whether these advantages are discernible or not to the average rider, the modification will be found on most second-hand models, and will have been carried out in any form from simply disconnecting the oil feed lines and pump cable to the complete removal of the system. To gain the maximum benefit the complete system should be removed as follows.

3 Working as described in the relevant Sections of Chapters 1 and 2, drain the oil tank, disconnect the tank/pump feed pipe and remove the tank complete with mountings, breather hose, and any clamps or ties securing the feed pipe. The oil pump cover should be removed and the feed pipes should be disconnected from the pump and the carburettor or intake stub, and pulled through the grommet set in the top of the crankcase right-hand cover. Disconnect the oil pump cable from the pump pulley, remove the adjuster, then remove the complete throttle/oil pump cable from the machine.

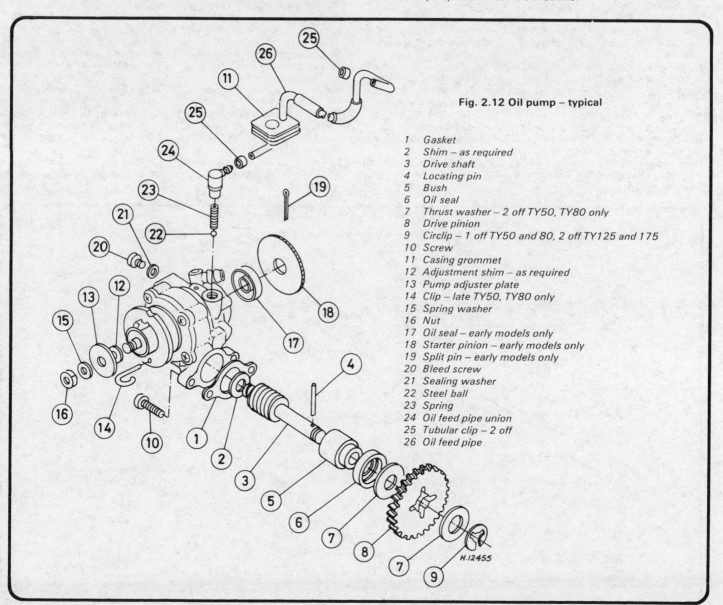

Fig. 2.12 Oil pump – typical

1 Gasket
2 Shim – as required
3 Drive shaft
4 Locating pin
5 Bush
6 Oil seal
7 Thrust washer – 2 off TY50, TY80 only
8 Drive pinion
9 Circlip – 1 off TY50 and 80, 2 off TY125 and 175
10 Screw
11 Casing grommet
12 Adjustment shim – as required
13 Pump adjuster plate
14 Clip – late TY50, TY80 only
15 Spring washer
16 Nut
17 Oil seal – early models only
18 Starter pinion – early models only
19 Split pin – early models only
20 Bleed screw
21 Sealing washer
22 Steel ball
23 Spring
24 Oil feed pipe union
25 Tubular clip – 2 off
26 Oil feed pipe

4   Substitute a single throttle cable (available from most specialist trials dealers) and adjust it as described in Routine Maintenance, then plug the oil feed pipe union on the carburettor or intake stub; the best method of doing this is to cut a thread inside the union and insert a small screw (8BA) secured with thread locking compound. If available, a small cap could be fitted over the union, but there is a risk that it might fall off and upset the carburation.

5   Remove the oil pump from the crankcase right-hand cover, then withdraw the cover itself and remove the oil pump drive pinion, shaft, oil seal and bush, as described in the previous Section. Using the pump gasket as a template, fabricate a blanking plate from a scrap piece of alloy sheet and fit it with the pump gasket over the driveshaft aperture using jointing compound if necessary to ensure an oil tight seal. If re-used, the pump mounting screws must be shortened to stop them from bottoming in their threaded holes, and thread locking compound employed to secure them. Refit the crankcase right-hand cover. The feed pipe grommet and oil pump cover can be refitted, if required.

6   Ensure that all parts of the conversion are reversible; you may wish to revert to Autolube lubrication at some future date. Carefully clean and store all components for this reason.

7   Consult a specialist dealer or expert rider for advice on what type of oil to use in the petroil and at what ratio. The traditional mineral-based self-mixing two-stroke oils are usually mixed at a ratio of 20 – 25:1, equivalent to 2/5 – 1/3 pint (230 – 180 cc) of oil mixed with one gallon of petrol; modern synthetic oils however are usually mixed at much leaner ratios (approx 50:1). Always use a self-mixing two-stroke oil, ordinary engine oil may be used in emergency only.

8   Apart from the need for more frequent decarbonisation due to the less efficient method of lubrication, there are several considerations to be borne in mind when converting to petroil lubrication. These are as follows:

9   The carburettor jets may require an increase in size of 5 – 10 per cent to allow for the passage of the oil, or the engine will be running weak at all throttle openings. Check this very carefully using the spark plug condition as a guide, (see colour photographs in Routine Maintenance) and jet up if necessary.

10 Never coast the machine down a long hill with the throttle closed, or the engine will seize from lack of lubricant; if the petrol supply is cut off, so is the lubricating oil.

11 Always carry a brand-new spare spark plug of the correct type and heat range with you; the likelihood of plug fouling or whiskering is much increased.

12 Always mix the fuel in a separate can and shake thoroughly before filling the tank. Also always switch the fuel tap off before filling the tank; if neat oil gets into the fuel pipe it will clog the carburettor jets. If the machine has been left standing for a long period of time, shake it vigorously to ensure that any oil is re-mixed that might have settled to the bottom of the tank.

13 Develop the habit of switching the fuel tap off before coming to a halt so that all the fuel in the carburettor and feed pipe is used up. If this is not done, the heat of the carburettor and engine will evaporate the petrol, leaving only neat oil to be forced into the jets when the fuel tap is next opened. This will clog the jets, necessitating the dismantling of the carburettror.

14 This is **not** a factory recommended conversion and will invalidate the warranty on a new machine.

13 Exhaust system: general

1   Check at regular intervals that the exhaust is securely fastened and that there are no leaks at any of the joints: leaks caused by damaged gaskets must be cured by the renewal of the gasket. Check that the system components are intact and have not been rotted away by corrosion; the only cure is the renewal of the affected component.

2   The matt-black painted finish is cheaper to renovate but less durable than a conventional chrome-plated system. It is inevitable that the original finish will deteriorate to the point where the system must be removed from the machine and repainted. Reference to the advertisements in the national motorcycle press, or to a local Yamaha agent and to the owners of machines with similarly-finished exhausts will help in selecting the most effective finish. The best are those which require the paint to be baked on, although some aerosol sprays are almost as effective.

3   Do not at any time attempt to modify the exhaust system in any way. The exhaust system is designed to give the maximum power possible consistent with legal requirements and yet to produce the minimum noise level possible. If an aftermarket accessory system is being considered, check very carefully that it will maintain or increase performance when compared with the standard system, without making excessive noise.

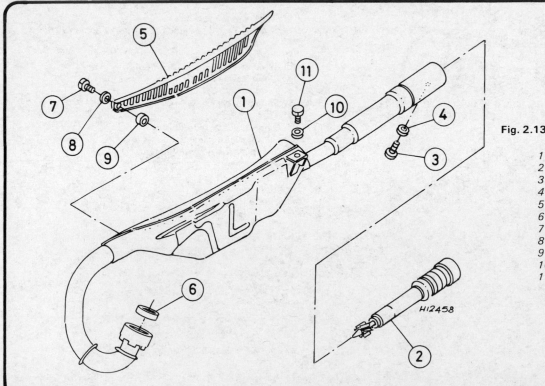

Fig. 2.13 Exhaust system – TY80

1 Exhaust system
2 Baffle
3 Screw
4 Spring washer
5 Heat shield
6 Gasket
7 Screw – 3 off
8 Washer – 3 off
9 Washer – 3 off
10 Washer
11 Bolt

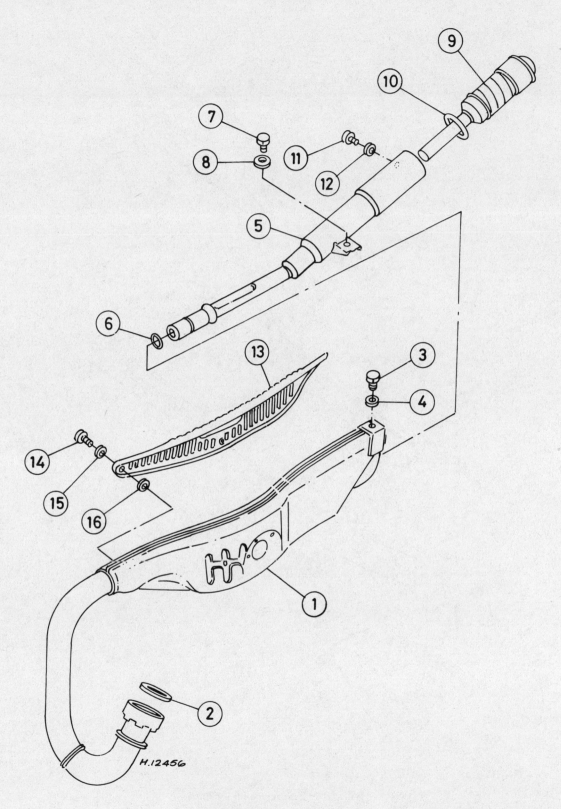

H.12456

Fig. 2.14 Exhaust system – TY50

1	Exhaust pipe	5	Silencer	9	Baffle	13	Heat shield
2	Gasket	6	O-ring	10	O-ring	14	Screw – 3 off
3	Bolt	7	Bolt	11	Screw	15	Washer – 3 off
4	Washer	8	Washer	12	Spring washer	16	Washer – 3 off

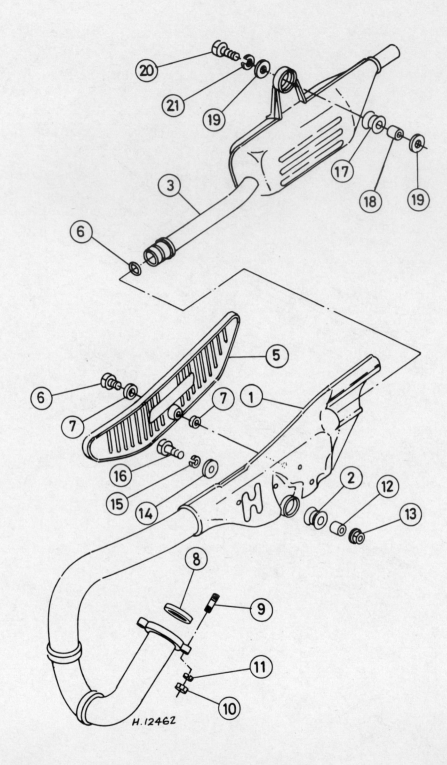

Fig. 2.15 Exhaust system – TY125

1 Exhaust pipe	7 Washer – 8 off	12 Spacer	17 Mounting rubber
2 Mounting rubber	8 Gasket	13 Nut	18 Spacer
3 Silencer	9 Stud – 2 off	14 Washer	19 Washer – 2 off
4 O-ring	10 Nut – 2 off	15 Spring washer	20 Bolt
5 Heat shield	11 Spring washer – 2 off	16 Bolt	21 Spring washer
6 Screw – 4 off			

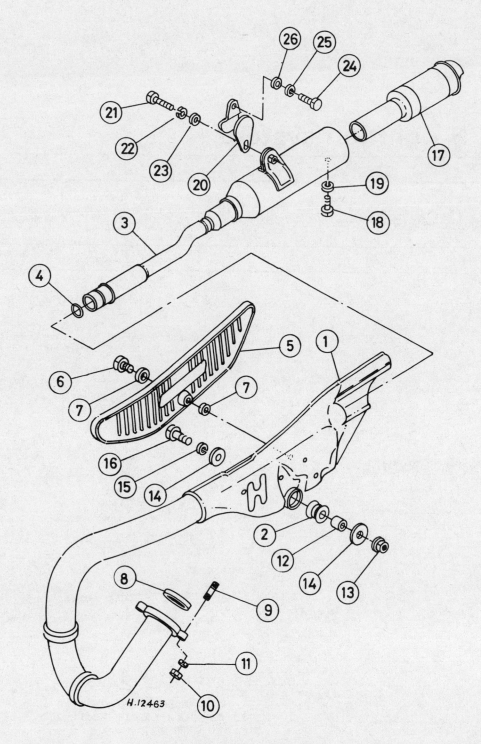

Fig. 2.16 Exhaust system – TY175

1 Exhaust pipe	8 Gasket	15 Spring washer	21 Bolt
2 Mounting rubber	9 Stud – 2 off	16 Bolt	22 Spring washer
3 Silencer	10 Nut – 2 off	17 Baffle	23 Washer
4 O-ring	11 Spring washer – 2 off	18 Screw	24 Bolt
5 Heat shield	12 Spacer	19 Spring washer	25 Spring washer
6 Screw – 4 off	13 Nut	20 Rear mounting	26 Washer
7 Washer – 8 off	14 Washer – 2 off		

Chapter 3 Ignition system

Contents

Specifications

Ignition timing
Piston position BTDC .... 1.80 ± 0.15 mm (0.0709 ± 0.0059 in)

Contact breaker
Gap .... 0.35 mm (0.014 in)
Tolerance .... 0.30 – 0.40 mm (0.012 – 0.016 in)

Ignition source coil
Winding resistance – black/earth:
   TY50 .... 1.35 ohm ± 10% @ 20°C (68°F)
   TY80 .... 1.50 ohm ± 10% @ 20°C (68°F)
   TY125, TY175 .... 1.60 ohm ± 10% @ 20°C (68°F)

Ignition HT coil
Minimum spark gap – at cranking speed:
   TY50, TY80, TY125 .... 6.0 mm (0.24 in)
   TY175 .... 7.0 mm (0.28 in)
Primary winding resistance:
   TY50 .... 1.02 ohm ± 10% @ 20°C (68°F)
   TY80 .... 1.70 ohm ± 10% @ 20°C (68°F)
   TY125, TY175 .... 4.50 ohm ± 10% @ 20°C (68°F)
Secondary winding resistance .... 6.00 K ohm ± 20% @ 20°C (68°F)

Condenser
Minimum resistance .... 3 M ohm
Capacity .... 0.30 microfarad ± 10%

Spark plug
Make .... NGK
Type:
   TY50 .... B7HS
   TY80 .... B6HS
   TY125, TY175 .... B7ES
Gap:
   TY50, TY80, TY175 .... 0.5 – 0.6 mm (0.020 – 0.024 in)
   TY125 .... 0.7 – 0.8 mm (0.028 – 0.032 in)
Plug cap resistance – TY50 .... 5.0 K ohm @ 20°C (68°F)

Torque wrench settings

Component	kgf m	lbf ft
Spark plug:		
TY50, TY80	2.5 – 3.0	18.0 – 22.0
TY125, TY175	2.0	14.5
Generator rotor nut:		
TY50	5.0 – 7.0	36.0 – 50.5
TY80	3.5 – 4.0	25.0 – 29.0
TY125, TY175	6.0 – 7.0	43.0 – 50.5

1 General description

A conventional contact breaker ignition system is fitted. As the generator rotor moves, alternating current (ac) is generated in the ignition source coil of the stator. With the contact breaker closed, the current runs to earth. When the breaker opens, the current transfers to the ignition coil primary windings. A high voltage is thus produced in the coil secondary windings (by mutual induction) and fed to the spark plug via the HT lead.

As energy flows to earth across the plug electrodes, a spark is produced and the combustible gases in the cylinder ignited. A condenser prevents arcing across the contact breaker points which helps reduce erosion due to burning.

This Chapter is concerned only with the testing and repair of the system; refer to Routine Maintenance for details of the regular attention needed for the spark plug, contact breaker points and ignition timing. Refer to the relevant Sections of Chapter 6 for details of how to check the wiring and switches.

2 Condenser (capacitor): testing and renewal

1 The condenser prevents arcing across the contact breaker points as they separate; it is connected in parallel with the points. If misfiring occurs or starting proves difficult especially with the engine hot it is

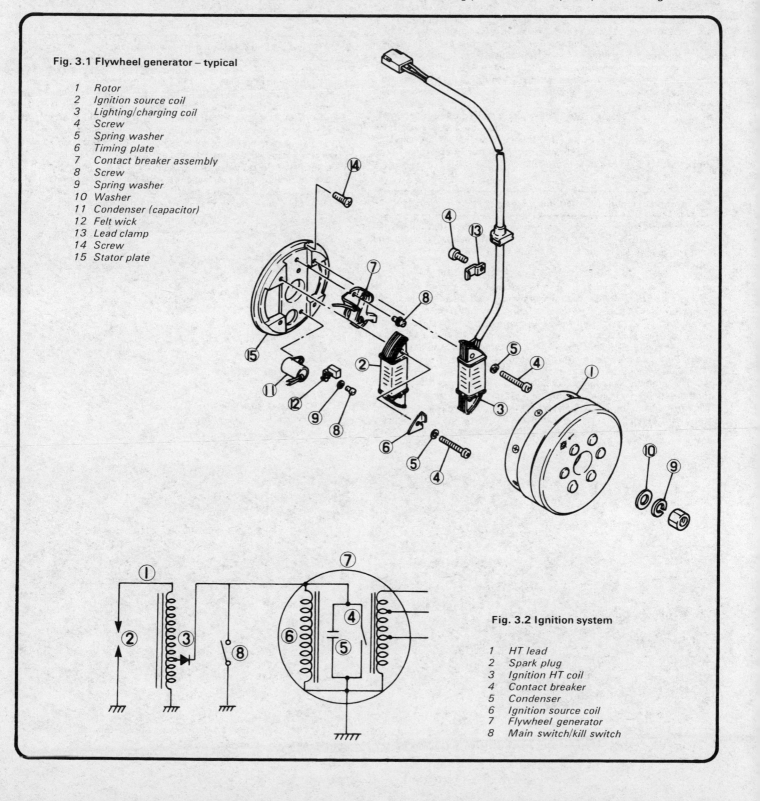

Fig. 3.1 Flywheel generator – typical

1 Rotor
2 Ignition source coil
3 Lighting/charging coil
4 Screw
5 Spring washer
6 Timing plate
7 Contact breaker assembly
8 Screw
9 Spring washer
10 Washer
11 Condenser (capacitor)
12 Felt wick
13 Lead clamp
14 Screw
15 Stator plate

Fig. 3.2 Ignition system

1 HT lead
2 Spark plug
3 Ignition HT coil
4 Contact breaker
5 Condenser
6 Ignition source coil
7 Flywheel generator
8 Main switch/kill switch

possible that the condenser is faulty. To check, watch the points via the rotor apertures when the engine is running. If the points spark excessively and appear burnt, the condenser is unserviceable.

2   The condenser can be tested in place, after the generator rotor has been removed as described in Chapter 1, but its leads must first be disconnected and an electro-tester (Yamaha part number 90890-03021) used, although a good quality ohmmeter can also be used for the first part of the test.

3   Using the tester or ohmmeter, measure the resistance between the condenser centre terminal and its outer casing; a value of at least 3 M ohms should be obtained (ie very close to infinite resistance).

4   Switch the tester to the condenser capacity test scale and connect its leads as described in the preceding paragraph. The needle should deflect and then return as the condenser is charged; note the reading obtained when the needle stops. A value of 0.3 microfarad should be obtained. After this test the condenser must be discharged by touching a length of thick insulated wire to its centre terminal and outer casing.

5   If the results of either test are not satisfactory, the condenser is faulty and must be renewed. If the test equipment is not available, the checks in paragraph 1 above must be assumed to be sufficiently accurate; if the symptoms described are found, the condenser should be renewed.

6   Remove the stator plate and unscrew the condenser mounting screw, then press out the condenser, once its leads have been unsoldered. On refitting ensure that the leads are soldered securely to the centre terminal and that they do not touch the outer casing. Check that all screws are securely fastened.

3   Ignition HT coil: location and testing

1   The ignition HT coil is mounted on the frame top tubes and requires no attention other than to ensure its mounting bolts and connections are clean and tight. If found to be faulty, the coil must be renewed as it is a sealed unit which cannot be repaired.

2   The most accurate test of a coil of this type is to check its performance on a spark-gap tester; a sound coil should produce a strong blue spark across a certain minimum gap (see Specifications) for at least 5 minutes. This can be approximated as a quick check when tracing faults by removing the suppressor cap and holding the end of the HT lead approximately $\frac{1}{4}$ in from the cylinder head finning while turning the engine over on the kickstart. Be very careful; the spark may find it easier to travel to earth by way of your hand, thus producing a most unpleasant shock.

3   The condition of the coil windings can be assessed by using a multimeter or ohmmeter set to the x 1 ohm scale for the primary windings and to the kilo ohms scale for the secondary windings. Make the meter connections for each test as shown in the accompanying illustration. If the readings obtained differ markedly from those given

in the Specifications section, the coil must be taken to a Yamaha Service Agent for testing as described in paragraph 2.

4   While all models are fitted with a diode to prevent the engine from running backwards, only on the early TY50 models is the diode separate from the coil, fitted in its low-tension lead. Check that current can flow in one direction only, if resistance is encountered in both directions, or in none at all, the diode must be renewed.

4   Ignition source coil: testing

The ignition source coil is tested exactly as described in Section 6 of Chapter 6, the reading expected from a sound coil being given in the Specifications Section of this Chapter.

5   HT lead and suppressor cap: examination

1   Erratic running faults can sometimes be attributed to leakage from the HT lead and spark plug. With this fault, it will often be possible to see tiny sparks around the lead and suppressor cap at night. One cause is dampness and the accumulation of road salts around the lead. It is often possible to cure the problem by drying and cleaning the components and spraying them with an aerosol ignition sealer.

2   If the system has become swamped with water, use a water dispersant spray. Renew the cap seals if defective. If the lead or cap is suspected of breaking down internally, renew the component.

3   Where the lead is permanently attached to the ignition coil, entrust its renewal to an auto-electrician who will have the expertise to solder on a new lead without damaging the coil windings.

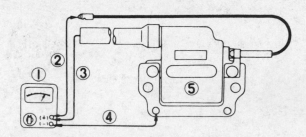

Fig. 3.3 Ignition HT coil resistance tests

1   Meter
2   Primary coil resistance value
3   Secondary coil resistance value
4   Earth
5   Ignition coil

2.6 Lift up lead retaining clamp and unsolder terminal to disconnect condenser

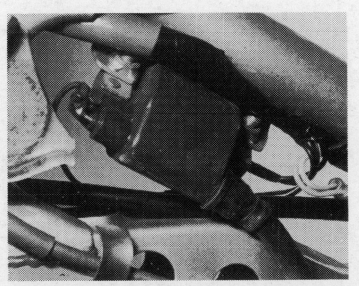

3.1 HT coil is mounted on frame top tubes

Chapter 4 Frame and forks

Contents

Specifications

Front forks

	TY50	TY80	TY125, TY175
Travel	110 mm (4.33 in)	100 mm (3.94 in)	160 mm (6.30 in)
Spring free length	227.0 mm (8.9370 in)	302.5 mm (11.9094 in)	418.5 mm (16.4765 in)
Fork oil capacity – per leg	119 cc (4.19 fl oz)	85 cc (2.99 fl oz)	126 cc (4.44 fl oz)
Recommended oil:			
TY50, TY125	SAE 10W/30SE engine oil, or fork oil		
TY80, TY175	SAE 10W, 20W or 30W fork oil		

Rear suspension

	TY50	TY80	TY125, TY175
Travel	75 mm (2.95 in)	65 mm (2.56 in)	105 mm (4.13 in)
Spring free length	191.5 mm (7.5394 in)	214.0 mm (8.4252 in)	224.5 mm (8.8386 in)
Maximum side-to-side play – at fork ends	1.0 mm (0.0394 in)	1.0 mm (0.0394 in)	1.0 mm (0.0394 in)

Torque wrench settings

Component	kgf m	lbf ft
Fork crown bolt (steering stem):		
TY50	5.0 – 8.0	36.0 – 58.0
TY80	3.5 – 4.0	25.0 – 29.0
TY125, TY175	4.2 – 6.5	30.0 – 47.0
Steering stem pinch bolt:		
TY50, TY80	N/App	N/App
TY125, TY175	0.8 – 1.25	6.0 – 9.0
Handlebar clamp bolts:		
TY50, TY80	1.5 – 2.5	11.0 – 18.0
TY125, TY175	1.0	7.0
Fork cap bolts:		
TY50	1.60 – 2.34	11.5 – 17.0
TY80, TY125, TY175	3.0 – 4.0	22.0 – 29.0
Top yoke pinch bolts:		
TY50	2.2 – 3.0	16.0 – 22.0
TY80	N/App	N/App
TY125, TY175	0.8 – 1.25	6.0 – 9.0
Bottom yoke pinch bolts:		
TY50, TY80	3.6 – 4.8	26.0 – 34.5
TY125, TY175	0.8 – 1.25	6.0 – 9.0
Swinging arm pivot bolt retaining nut:		
TY50, TY80	2.0 – 3.0	14.5 – 22.0
TY125, TY175	4.5	32.5
Suspension unit mountings:		
TY50	3.0 – 4.8	22.0 – 34.5
TY80	N/Av	N/Av
TY125, TY175	0.5	3.5
Rear brake torque arm/swinging arm retaining nut	1.5	11.0

1 General description

The front forks are telescopic, coil sprung, with hydraulic damping, while the rear suspension consists of two coil sprung hydraulically-damped suspension units acting on a pivoted fork (swinging arm). The frame is all welded tubular steel of the full cradle type.

2 Front fork legs: removal and refitting

1 Place the machine securely on a stout wooden box or paddock stand so that the front wheel is clear of the ground. The front wheel can then be removed according to the instructions given in the relevant Section of Chapter 5.

2 Remove the four mudguard mounting bolts then carefully withdraw the mudguard.

3 On TY80 models, the fork legs are retained in the top yoke by the fork top bolt. Remove the fork top bolts and the bottom yoke pinch bolts. It is then possible to slide the fork leg down and away from the yoke. On all other models, the fork legs are retained by pinch bolts in both top and bottom yokes. On these models slacken the pinch bolts and slide the fork legs down and out of the yokes. Note that the top bolts should be slackened first, if the forks are to be dismantled.

4 If the fork legs are stuck in the yokes, apply penetrating fluid and attempt to rotate the legs by hand to free them. It may be necessary to completely remove the pinch bolts and to spring the clamps apart slightly with a large, flat-bladed screwdriver. Great care must be taken not to distort or to break the clamp, as this will necessitate renewal of the complete yoke. If the leg is still reluctant to move, push a metal bar of suitable diameter through the spindle lug in the fork lower leg and tap firmly downwards on the protruding end of the bar to drive the fork leg from the yokes.

5 Once the legs have been removed, put them to one side to await stripping and examination. If they are not to be dismantled, ensure that they remain upright so that no fork oil is lost.

6 Reassembly is a straightforward reversal of the removal sequence, noting the following points. On TY80 models, the leg must be pushed up through the bottom yoke to the underside of the top yoke and retained there by firmly tightening the top bolt. On TY50 models push the fork leg up through both yokes until the top of the chromed stanchion tube is flush with the upper surface of the top yoke while on TY125 and TY175 models the top of the stanchion should project 20 mm (0.79 in) above the upper surface of the top yoke. Each stanchion must be at exactly the same height. Once in place, the fork legs should be retained by tightening the top yoke pinch bolts as lightly as possible. For both types of forks, refitting the legs will be made easier if a small amount of grease or oil is smeared over the upper length of the stanchion.

7 When fitting the front mudguard and wheel back into the forks, tighten the spindle nut and other mounting bolts only lightly at first and take the machine off its centre stand or box. Apply the front brake and push down on the handlebars several times so that the operation of the fork legs settles each component in its correct place. Using a torque wrench, tighten all nuts and bolts from the wheel spindle upwards to the fork top bolt or top yoke pinch bolt, to the torque settings given in the Specifications Section of this Chapter. This will ensure that the fork components can operate freely and easily, with no undue strain imposed from an overtightened bolt or an awkwardly positioned part.

8 Be very careful to check fork operation, front brake adjustment and that all nuts and bolts are securely fastened before taking the machine out on the road.

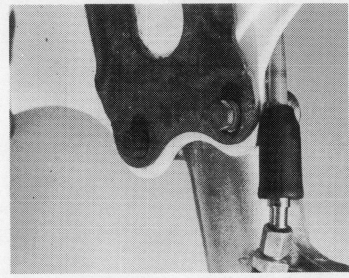

2.2 Remove mounting bolts to release front mudguard

2.3a Slacken top yoke pinch bolts (where fitted) ...

2.3b ... followed by bottom yoke pinch bolts ...

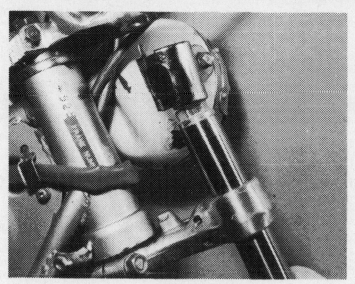

2.3c ... and slide fork leg downwards out of yokes

3 Front fork legs: dismantling and reassembly

1   Dismantle and rebuild the fork legs separately so that there is no chance of exchanging components, thus promoting undue wear.

TY50 and TY80 models
2   Remove the O-ring from the top of the fork stanchion and invert the leg over a container to drain the oil. Pump the leg several times to ensure that as much oil as possible is removed.
3   Displace the dust seal from the top of the fork lower leg and on late models only, withdraw the circlip and the plain washer from the top of the lower leg, above the oil seal. Using soft jaw covers to prevent the alloy from being marked, clamp the fork lower leg wheel spindle lug in a vice and withdraw the stanchion with the seal and bushes. To do this, press the stanchion in as far as possible, then withdraw it sharply. This 'slide-hammer' action must be repeated several times before the stanchion will be released. In especially stubborn cases, take the fork leg to a Yamaha Service Agent who can dismantle it using a slide hammer 90890-01290 and adaptor 90890-01291. For those owners who have the facilities, a slide hammer arrangement can be devised. With the oil seal released from the fork lower leg, withdraw the stanchion complete with the bushes. Remove from inside the stanchion the fork spring, noting carefully which way round it is fitted, the washer and the spacer. Invert the lower leg to tip out the damper rod.
4   On early models, identified by the chome-plated seal holder at the top of each fork lower leg, unscrew the seal holder without marking it. This can be done using a strap wrench with the lower leg spindle lug clamped in a vice, or by cutting a large V-shaped notch in two blocks of wood. Place the two blocks in a vice with the seal holder in the middle of the two notches and tighten the vice to form a large clamp. Using the spindle as a tommy bar in the spindle lug, unscrew the fork leg from the seal holder. With the seal holder removed, the seal can be driven out of it. Remove the stanchion and fork spring as described above and withdraw the O-ring from the top of the lower leg.
5   Carefully clean the upper length of the stanchion, removing any dirt or corrosion, smear oil along its length and slide off the oil seal (late models only) and the fork top bush. Note that since this will almost certainly damage the seal lips, in spite of the precautions described, the seals must be renewed whenever the forks are dismantled. On late TY80 models only the remaining bush can be removed after its retaining circlip has been withdrawn.
6   On reassembly, it is assumed that all components are scrupulously clean. On late TY80 models only, refit the bottom bush to the stanchion ensuring that all circlips are correctly seated and do not stand proud of their bushes. Oil the stanchion surface before refitting the top bush and a new oil seal (late models only).
7   Having refitting the spacer and washer inside the stanchion, use

the notes made on dismantling to refit the fork spring the correct way up; usually, the close-pitched coils are uppermost. If no notes were made, wear marks on the washer and on the damper rod might reveal which end of the spring bore against them. Insert the damper rod into the fork lower leg (where fitted).
8   Smear oil over the bushes and insert the stanchion fully into the lower leg, then press the top bush down so that its shoulder locates in the recess in the lower leg.
9   On late models, smear grease over the outside of the new fork seal and press it into the lower leg as far as possible by hand, ensuring that it enters squarely into its housing. Refit the plain washer on top of the seal which must be driven into the lower leg far enough to expose the circlip groove and no further. The Yamaha service tool is a metal tube about 2 – 3 inches long and of the same inside and outside diameter as the seal, with two handles welded to it so that pressure can be applied. If this is not available, a substitute can be fabricated; this is worthwhile if the forks are to be dismantled very often. Ensure that the surface which contacts the seal is free from burrs or sharp edges. The only alternative is to use a hammer and drift, tapping evenly all the way round the seal top surface; this is a method which requires care and patience if the seal is not to be damaged. Ensure that the drift bears only on the metal washer to avoid damage to the seal, and be very careful that the seal remains square in its housing. As soon as the circlip groove is fully exposed above the metal washer, refit the seal retaining circlip, ensuring that it is correctly engaged in the groove. Pack the space above the seal with grease as additional protection against dirt and corrosion then refit the dust seal. Pump the fork several times to ensure that it is working smoothly and easily.
10   On early models, check that the threads of the fork lower leg and seal holder are completely clean and undamaged, then fit a new O-ring to the top of the lower leg. Ensuring that it is fitted with its spring-loaded centre lip facing downwards, drift the new oil seal into the seal holder using a hammer and a tubular drift such as a socket spanner that bears only on the seal's hard outer edge. Be very careful not to damage the seal. Grease the seal lips and smear oil over the stanchion, then slide the seal holder slowly down the length of the stanchion, taking care not to damage the seal lips, and screw it into place on the fork lower leg. Tighten it securely and pack the space above the seal with grease, refit the dust seal and pump the fork several times to check its action.
11   Refill the leg with the correct amount of the recommended grade of oil. Due to the small size of the filler orifice oil will have to be added using a cheap syringe, as shown, or poured in very slowly through a funnel. Be careful to put exactly the same amount of oil in both legs. Refit the O-ring in the recess at the top of the stanchion.

TY125 and TY175 models
12   Unscrew the fork top bolt; as previously mentioned this is easiest while the fork leg is still in the machine as the alternative is to use soft alloy or wooden jaw covers to prevent the stanchion from being marked while it is held in a vice. Remove the top bolt with its O-ring, the spacer, washer and the fork spring. Make a careful note of which way up the fork spring is fitted. Invert the leg over a suitable container to drain the oil. Pump the leg several times to ensure that as much oil as possible is removed, then displace the dust seal from the top of the fork lower leg.
13   Carefully clamp the fork lower leg in a vice to avoid distortion and use an Allen key to unscrew the damper rod retaining bolt which is set in a recess in the base of the fork lower leg. In some cases the bolt will unscrew with ease, but it is more usual for the bolt to free itself from the lower leg and then rotate with the damper rod assembly, so that nothing useful is achieved. In such a case, obtain a length of metal tubing of the same diameter and wall thickness as the fork spring, which will fit closely over the head of the damper rod, and insert the tubing down into the bore of the stanchion. The services of an assistant will now be required. Clamp a self-locking wrench to the protruding end of the tubing and with the assistant preventing the tubing from turning and simultaneously applying pressure via the tubing to the head of the damper rod, the damper rod will be locked in place so that the retaining bolt can be unscrewed. When working alone, use a longer length of tubing which can be clamped in the vice. By pushing down on the fork lower leg with one hand it should be possible to lock the damper rod firmly enough for the retaining bolt to be unscrewed.
14   Carefully pull the stanchion out of the lower leg. Either invert the stanchion to tip out the damper rod and spring or remove the circlip at

the bottom of the stanchion and pull out the complete damper assembly. If this is to be dismantled, be very careful to note exactly the order in which parts are fitted and which way up each component is aligned. Note that there is little point in dismantling the damper assembly except for cleaning, as there are no individual components available with which it can be reconditioned. Invert the lower leg to tip out the damper rod seat.

15 The fork oil seal fitted to each leg should be renewed whenever the stanchion is removed and must be renewed if it is disturbed as the means used for removing it will almost certainly cause damage. The seal is retained by a circlip which must be removed using a small, flat-bladed screwdriver to ease it away from its groove in the fork lower leg followed by a plain washer. Use a large flat-bladed screwdriver to lever the seal from its housing. Take care not to scratch the internal surface of the seal housing with the edge of the screwdriver blade, and do not apply excessive pressure or there is a risk of the upper edge of the fork lower leg being cracked or distorted. If the seal appears difficult to move, heat the leg by pouring boiling water over its outer surface. This will cause the alloy leg to expand sufficiently to loosen the seal.

16 On reassembly, it is assumed that all components are scrupulously clean. Refer to the illustration accompanying the text for additional guidance. Clamp the lower leg securely in a vice by means of the wheel spindle boss. Coat the inner and outer diameters of the seal with the recommended fork oil and push the seal squarely into the bore of the fork lower leg by hand. Ensure that the seal is fitted squarely, then refit the plain washer and tap it fully into position, using a hammer and a suitably sized drift such as a socket spanner, which should bear only on the washer. Tap the seal into the bore of the lower leg just enough to expose the circlip groove. Refit the retaining circlip securely in its groove. Refit the damper rod assembly to the stanchion, using the accompanying photographs for guidance, then place the damper rod seat over the damper rod end, using a smear of grease to stick it in place.

17 Smear the sliding surface of the stanchion with a light coating of fork oil and carefully insert the stanchion into the lower leg, taking great care not to damage the sealing lips of the oil seal. Push a fork spring or the length of tubing used on dismantling into the stanchion and apply pressure on this to ensure that the damper rod or its seat is pressed firmly into the base of the lower leg. Check that the threads of the damper rod bolt are clean and dry, apply a few drops of thread locking compound and fit the damper rod bolt. Do not forget the sealing washer fitted under the head of the bolt. Tighten the bolt only partially at first, using an Allen key of suitable size. Maintain pressure on the head of the damper rod and push the stanchion firmly as far down into the lower leg as possible to centralise the damper rod in the stanchion. The damper rod bolt can then be tightened firmly. Withdraw the spring or tubing from the stanchion.

18 Pack the space above the seal with grease as additional protection against dirt or corrosion, then refit the dust seal. Fill the leg with 126 cc (4.44 fl oz) of the recommended fork oil, using a finely graduated measuring vessel to ensure that exactly the same amount of oil is put in each leg. Very slowly pump the leg to distribute the oil, then pull the stanchion out of the leg as far as possible and insert the fork spring using the notes made on dismantling to ensure that it is fitted the correct way up; usually, the closer-pitched coils are uppermost. Refit the washer, the spacer and the top bolt, complete with its O-ring. Tighten securely the top bolt to a torque setting of 3.0 – 4.0 kgf m (22.0 – 29.0 lbf ft).

4 Front fork legs: examination and renovation

1   Carefully clean and dry all the components of the fork leg. Lay them out on a clean work surface and inspect each one, looking for excessive wear, cracks, or other damage. All traces of oil, dirt, and swarf should be removed, and any damaged or worn components renewed.

2   Examine the sliding surface of the stanchion or bushes, as applicable, and the internal surface of the lower leg, looking for signs of scuffing which will indicate that excessive wear has taken place. Slide the stanchion into the lower leg so that it seats fully. Any wear present will be easily found by attempting to move the stanchion backwards and forwards, and from side to side, in the bore of the lower leg. It is inevitable that a certain degree of slackness will be found, especially when the test is repeated at different points as the stanchion is gradually withdrawn from the lower leg, and it is largely a matter of experience to assess with accuracy the amount of wear necessary to justify renewal of either the stanchion, the bushes, or the lower leg. It is recommended that the two components be taken to a motorcycle dealer for an expert opinion to be given if there is any doubt about the degree of wear found. Note that while wear will only become a serious problem after a high mileage has been covered, it is essential that such wear be rectified by the renewal of the components concerned if the handling and stability of the machine are impaired.

3   Check the outer surface of the stanchion for scratches or roughness; it is only too easy to damage the oil seal during reassembly if these high spots are not eased down. The stanchions are unlikely to bend unless the machine is damaged in an accident. Any significant bend will be detected by eye, but if there is any doubt about straightness, roll down the stanchion tubes on a flat surface such as a sheet of plate glass. If the stanchions are bent, they must be renewed. Unless specialised repair equipment is available it is rarely practicable to effect a satisfactory repair.

4   Check the stanchion sliding surface for pits caused by corrosion. Such pits should be smoothed down with fine emery paper and filled,

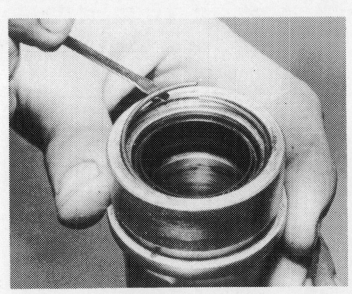

3.15a Lever out seal retaining circlip ...

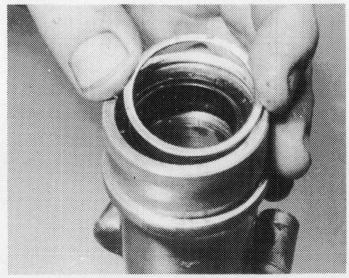

3.15b ... followed by the plain washer

if necessary, with Araldite. Once the Araldite has set fully hard, use a fine file or emery paper to rub it down so that the original contour of the stanchion is restored. Damage of this nature can be eliminated by the fitting of gaiters; these are available from any good motorcycle dealer.

5   After an extended period of service, the fork springs may take a permanent set. If the spring lengths are suspect, then they should be measured and the readings obtained compared with the lengths given in the Specifications Section of this Chapter. It is always advisable to fit new fork springs where the length of the original items has decreased by a significant amount. Always renew the springs as a set, never separately.

6   On TY125 and TY175 models if oil changes have not improved the situation, the complete damper assembly must be renewed if the fork performance has deteriorated. As previously mentioned Yamaha do not supply individual replacement parts.

7   Closely examine the dust seal for splits or signs of deterioration. If found to be defective, it must be renewed as any ingress of dirt will rapidly accelerate wear of the oil seal and fork stanchion. It is advisable to renew any gasket washers fitted beneath bolt heads as a matter of course. The same applies to the O-rings fitted to the fork top bolts and especially to the fork seals themselves.

3.15c Seal is levered from lower leg as shown – take care not to damage the leg

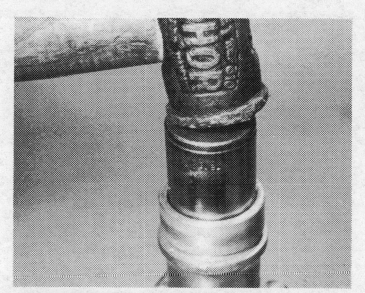

3.16a Care is required when fitting new fork seals

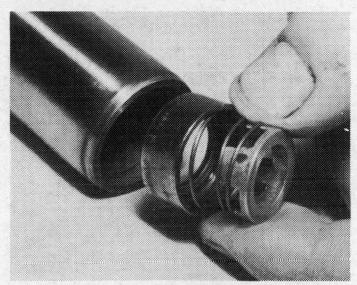

3.16b Damper valve assembly fits in stanchion lower end as shown ...

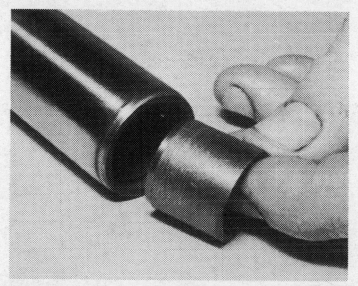

3.16c ... and is followed by damper piston

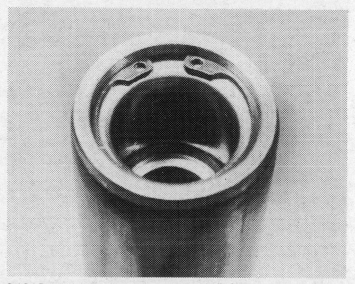

3.16d Damper assembly is secured by a circlip in stanchion – damper rod shown removed for clarity

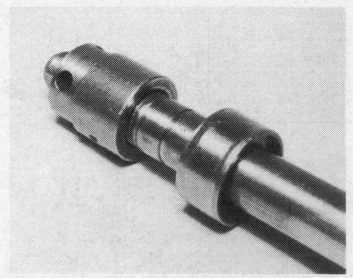
3.16e Valve block is fitted under damper rod head ...

3.16f ... and is secured by a circlip

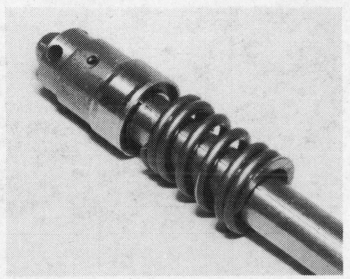
3.16g Do not omit rebound spring from damper assembly

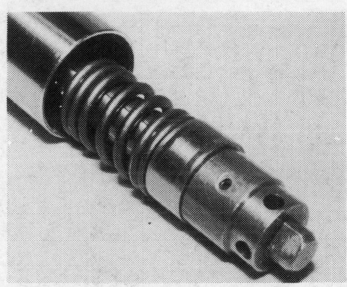
3.16h Damper rod can be inserted in stanchion upper end

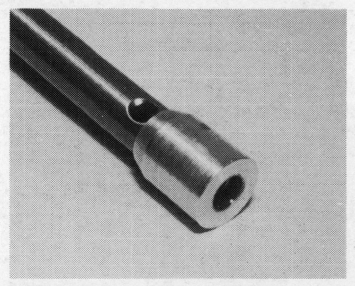
3.16i Fit damper rod seat to protruding end of damper rod

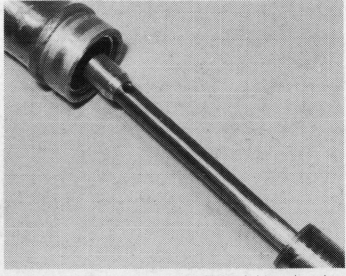
3.17a Insert stanchion and damper assembly into fork lower leg – do not damage seal

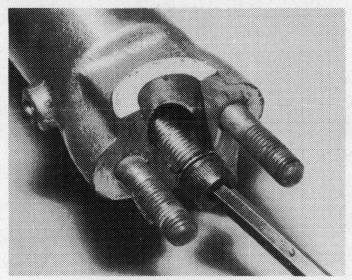

3.17b Tighten securely damper rod retaining bolt

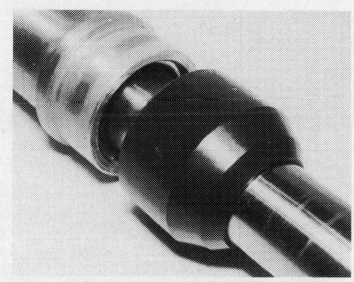

3.18a Pack grease above seal before refitting dust seal

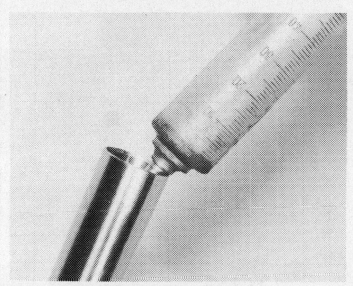

3.18b Refill each fork leg with exactly the correct amount of fork oil

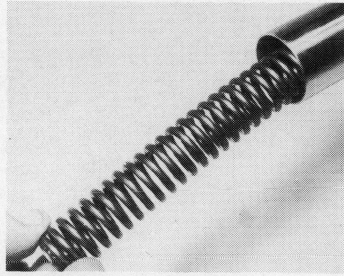

3.18c Fork spring closer-pitched coils are fitted upwards

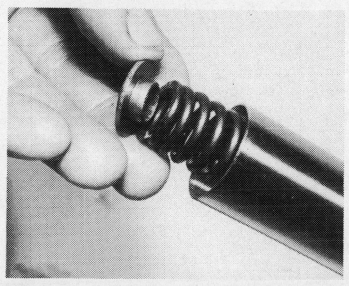

3.18d Washer forming spring top seat is fitted as shown ...

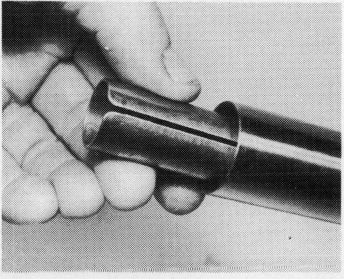

3.18e ... and is followed by the spacer

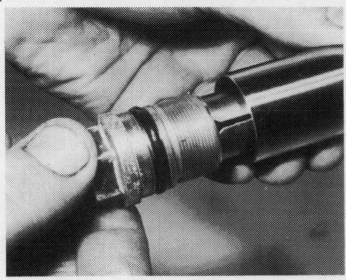

3.18f Renew O-ring around fork top bolt if worn or damaged

Fig. 4.1 Front forks – TY50 P and early TY50 M

1 Right-hand fork leg
2 Left-hand lower leg
3 Right-hand lower leg
4 O-ring
5 Spring
6 Spring seat
7 Top bush
8 Oil seal holder
9 Oil seal
10 Dust seal
11 Stanchion
12 O-ring
13 Washer
14 Top bolt
15 Steering stem
16 Bolt
17 Cable guide
18 Bolt
19 Left-hand headlamp bracket
20 Right-hand headlamp bracket
21 Screw – 2 off
22 Spring washer – 2 off

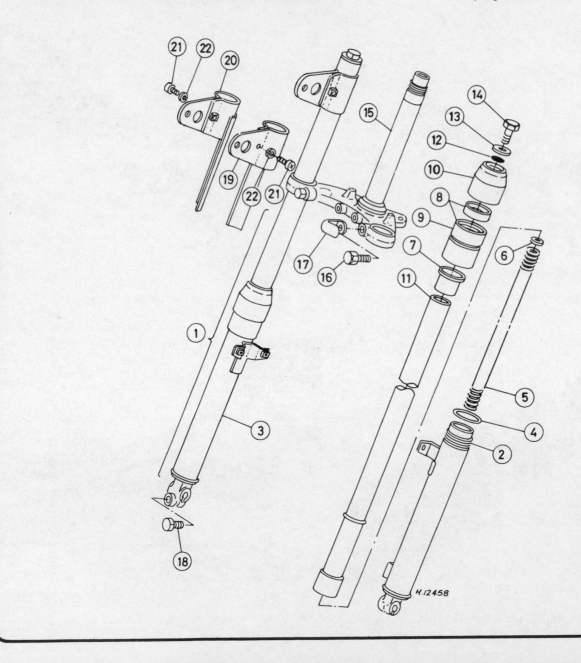

H.12458

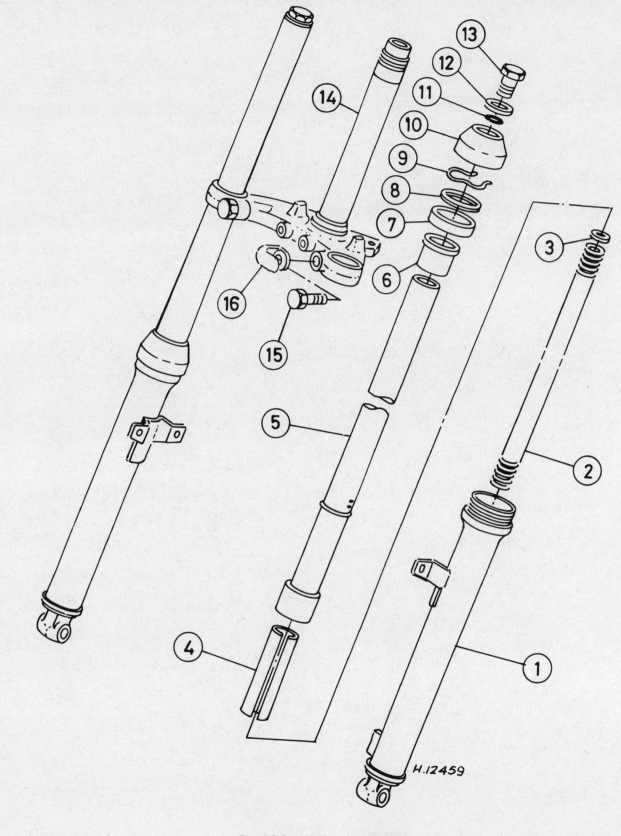

H.12459

Fig. 4.2 Front forks – late TY50 M

1 Lower leg	5 Stanchion	9 Circlip	13 Top bolt
2 Spring	6 Top bush	10 Dust seal	14 Steering stem
3 Spring seat	7 Oil seal	11 O-ring	15 Bolt
4 Spacer	8 Washer	12 Washer	16 Cable guide

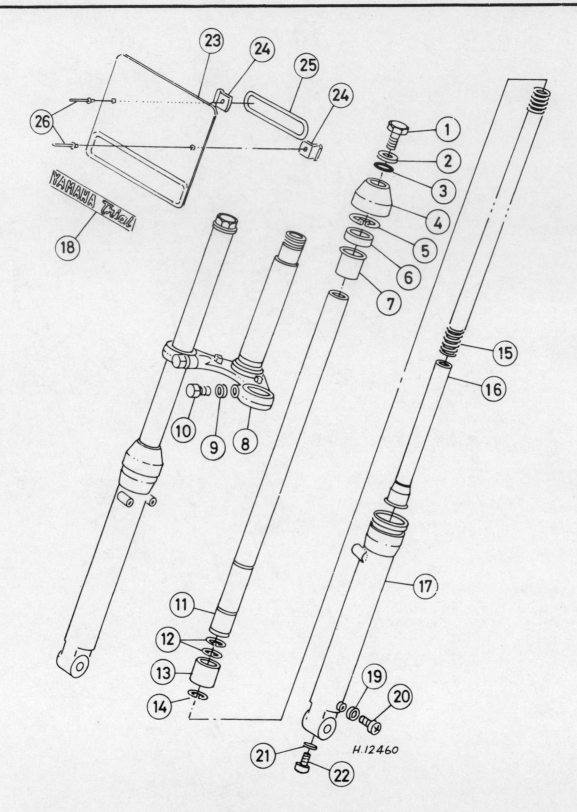

Fig. 4.3 Front forks – TY80

1 Top bolt	8 Steering stem	15 Spring	21 Sealing washer
2 Washer	9 Spring washer	16 Damper rod	22 Screw
3 O-ring	10 Bolt	17 Lower leg	23 Number plate
4 Dust seal	11 Stanchion	18 Emblem	24 Clip
5 Circlip	12 Circlip	19 Sealing washer	25 Rubber band
6 Oil seal	13 Bottom bush	20 Drain screw	26 Rivet
7 Top bush	14 Circlip		

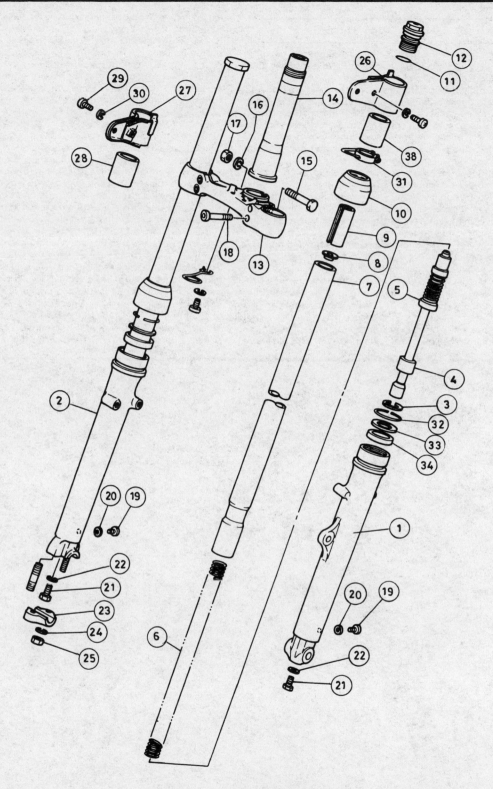

Fig. 4.4 Front forks – TY125 and 175

1	Left-hand lower leg	10	Dust seal	19	Drain plug	27	Right-hand headlamp bracket
2	Right-hand lower leg	11	O-ring	20	Gasket	28	Damping rubber
3	Circlip	12	Cap bolt	21	Bolt	29	Screw
4	Piston	13	Bottom yoke	22	Gasket	30	Spring washer
5	Damper rod	14	Steering stem	23	Spindle clamp	31	Cable stay
6	Spring	15	Bolt	24	Washer	32	Circlip
7	Stanchion	16	Spring washer	25	Nut	33	Washer
8	Spring guide	17	Nut	26	Left hand headlamp bracket	34	Oil seal
9	Spacer	18	Bolt				

5 Steering head assembly: removal and refitting

1   Working as described in Section 2 of this Chapter, remove the front wheel, the mudguard (where applicable) and the front forks, then remove the seat, the side panel and the fuel tank.

2   Remove the four handlebar clamp bolts and the handlebar clamps. Move the handlebars backwards as far as possible without straining the control cables or wiring and secure them clear of the steering head area. Remove or disconnect all control cables, wiring and the speedometer cable (where applicable) which might hinder the removal of the fork yokes. Components such as the horn and speedometer (where applicable) can remain in place on the yokes.

3   Remove the large chromium-plated bolt from its location through the centre of the top yoke. Using a soft-faced hammer, give the top yoke a gentle tap to free it from the steering head and lift it from position. Carry out a final check around the bottom yoke to ensure that all components have been removed which might prevent its release.

4   Support the weight of the bottom yoke and, using a C spanner, of the correct size, remove the steering head bearing adjusting ring. If a C-spanner is not available, a soft metal drift may be used in conjunction with a hammer to slacken the ring.

5   Remove the dust excluder and the cone of the upper bearing. The bottom yoke, complete with steering stem, can now be lowered from position. Ensure that any balls that fall from the bearings as the bearing races separate are caught and retained. It is quite likely that only the balls from the lower bearing will drop free, since those of the upper bearing will remain seated in the bearing cup. Full details of

examining and renovating the steering head bearings are given in Section 6 of this Chapter.

6   Fitting of the steering head assembly is a direct reversal of that procedure used for removal, whilst taking into account the following points. It is advisable to position all nineteen balls of the lower bearing around the bearing cone before inserting the steering stem fully into the steering head. Retain these balls in position with grease of the recommended type and fill both bearing cups with the same type of grease.

7   With the bottom yoke pressed fully home into the steering head, place the twenty-two balls into the upper bearing cup and fit the bearing cone followed by the dust excluder. Refit the adjusting ring and tighten it, finger-tight. The ring should now be tightened firmly to seat the bearings, using a C-spanner only; do not apply excessive force. Turn the bottom yoke from lock to lock five or six times to settle the balls, then slacken the adjusting ring until all pressure is removed.

8   To provide the initial setting for steering head bearing adjustment, tighten the adjusting ring carefully until resistance is felt then loosen it by $\frac{1}{8}$ to $\frac{1}{4}$ of a turn. Remember to check that the adjustment is correct, as described in Routine Maintenance, when the steering head assembly has been reassembled and the forks and front wheel refitted.

9   Finally, whilst refitting and reconnecting all disturbed components, take care to ensure that all control cables, drive cables, electrical leads, etc are correctly routed and that reference is made to the list of torque wrench settings given in the Specifications Section of this Chapter and of Chapter 5. Check that the headlamp beam height has not been disturbed and ensure that all controls and instruments function correctly before taking the machine on the public highway.

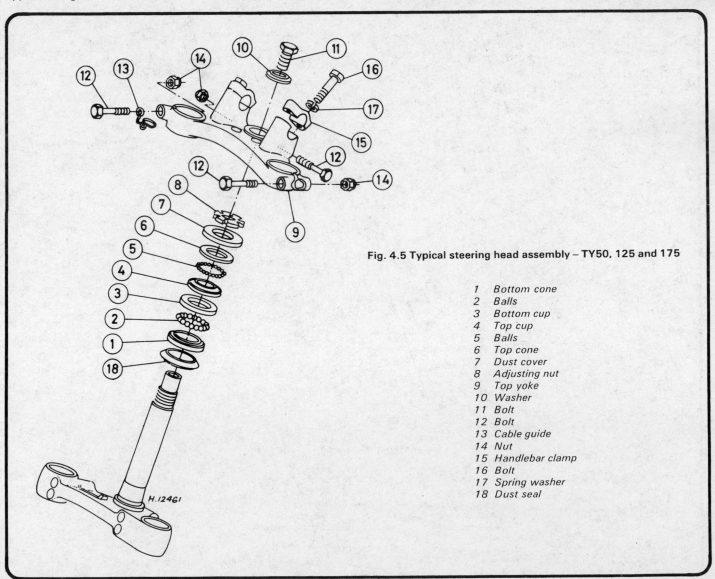

Fig. 4.5 Typical steering head assembly – TY50, 125 and 175

1   Bottom cone
2   Balls
3   Bottom cup
4   Top cup
5   Balls
6   Top cone
7   Dust cover
8   Adjusting nut
9   Top yoke
10 Washer
11 Bolt
12 Bolt
13 Cable guide
14 Nut
15 Handlebar clamp
16 Bolt
17 Spring washer
18 Dust seal

H.12461

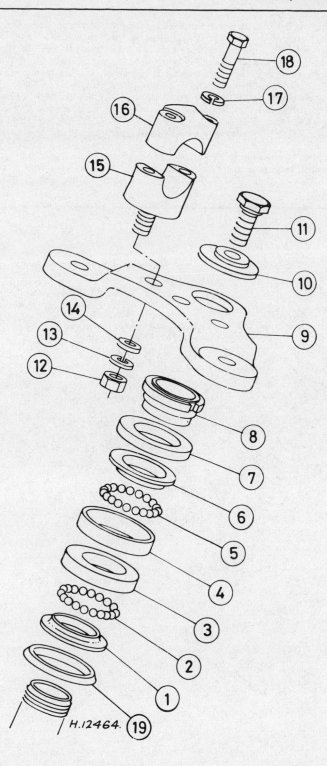

Fig. 4.6 Steering head – TY80

1	Bottom cone	11	Bolt
2	Balls	12	Nut
3	Bottom cup	13	Spring washer
4	Top cup	14	Washer
5	Balls	15	Handlebar lower clamp
6	Top cone	16	Handlebar upper clamp
7	Dust cover	17	Spring washer
8	Adjusting nut	18	Bolt
9	Top yoke	19	Dust seal
10	Washer		

6 Steering head bearings: examination and renovation

1 Before commencing reassembly of the steering head component parts, take care to examine each of the steering head bearings. The ball bearing tracks of their respective cup and cone bearings should be polished and free from any indentations or cracks. If wear or damage is evident, then the cups and cones must be renewed as complete sets.

2 Carefully clean and examine the balls contained in each bearing assembly. These should also be polished and show no signs of surface cracks or blemishes. If any one ball is found to be defective, then the complete set should be renewed. Remember that a complete set of these balls is relatively cheap and it is not worth the risk of refitting items that are in doubtful condition. Note that two different sizes of ball are used, the upper bearing fitted with $\frac{3}{16}$ in balls and the lower bearing with $\frac{1}{4}$ in items. Be careful at all times not to confuse the two sizes and ensure that each bearing is fitted with balls of the correct size.

3 Twenty-two balls are fitted in the top bearing and nineteen in the lower. This arrangement will leave a gap between any two balls but an extra ball must not be fitted, otherwise the balls will press against each other thereby accelerating wear and causing the steering action to be stiff.

4 The bearing cups are a drive fit in the steering head and may be removed by passing a long drift through the inner bore of the steering head and drifting out the defective item from the opposite end. The drift must be moved progressively around the race to ensure that it leaves the steering head evenly and squarely.

5 The lower of the two cones fits over the steering stem and may be removed by carefully drifting it up the length of the stem with a flat-ended chisel or a similar tool. Again, take care to ensure that the cone is kept square to the stem. There is a rubber seal fitted beneath this cone on some models; the seal must be renewed if damaged or worn to prevent the entry of dirt into the bearing.

6 Fitting of the new cups and cone is a straightforward procedure whilst taking note of the following points. Ensure that the cup locations within the steering head are clean and free of rust; the same applies to the steering stem. Lightly grease the stem and head locations to aid fitting and drift each cup or cone into position whilst keeping it square to its location. Fitting of the cups into the steering head will be made easier if the opposite end of the head to which the cup is being fitted has a wooden block placed against it to absorb some of the shock as the drift strikes the cup.

7 Frame: examination and renovation

1 The frame is unlikely to require attention unless accident damage has occurred. In some cases, renewal of the frame is the only satisfactory remedy if the frame is badly out of alignment. Only a few frame specialists have the jigs and mandrels necessary for resetting the frame to the required standard of accuracy, and even then there is no easy means of assessing to what extent the frame may have been overstressed.

2 After the machine has covered a considerable mileage, it is advisable to examine the frame closely for signs of cracking or splitting at the welded joints. Rust corrosion can also cause weakness at these joints. Minor damage can be repaired by welding or brazing, depending on the extent and nature of the damage.

3 Remember that a frame which is out of alignment will cause handling problems and may even promote 'speed wobble'. If misalignment is suspected, as a result of an accident, it will be necessary to strip the machine completely so that the frame can be checked, and if necessary, renewed.

8 Swinging arm: removal and refitting

1 Remove the rear wheel following the instructions given in the relevant Section of Chapter 5.

2 Remove, if required, the brake torque arm and the chainguard from the swinging arm, then slacken all four suspension unit mounting nuts or screws. Remove the two suspension unit bottom mounting

nuts or screws and their washers, then pull the units sideways off their swinging arm mounting lugs. Note carefully the position and number of the plain washers at each mounting.

3   Remove the swinging arm pivot bolt securing nut, then withdraw the pivot bolt. If the bolt proves stubborn, apply a good quantity of penetrating fluid, allow time for it to work, then displace the bolt using a hammer and a metal drift. Withdraw the swinging arm. On TY125 and TY175 models it is not unknown for the inner sleeve to seize solidly on the pivot bolt with rust; in such cases the only remedy is to hacksaw through the bolt between each side of the swinging arm and the frame. The swinging arm can then be removed for the remains of the sleeve and shaft to be hammered out and renewed. Regular dismantling and lubrication, as described in Routine Maintenance, would ensure that this problem never arises.

4   Reassembly is the reverse of the dismantling procedure. Check that the pivot bolt and the passages through which it fits are completely clean and free from corrosion, then apply liberal quantities of grease to the surface of the bolt and to the passages, not forgetting the inside diameter of the bushes themselves. Offer up the swinging arm and insert the pivot bolt, then refit the retaining nut. Refit the suspension units, ensuring that the plain washers are correctly refitted, and tighten all four suspension unit mounting nuts or screws to the specified torque setting, then tighten the pivot bolt retaining nut to its specified torque setting. Complete the remainder of the reassembly work.

5   Before taking the machine out on the road, check that all the nuts and bolts are securely fastened, and that the rear suspension, rear brake, and chain tension are adjusted correctly and working properly.

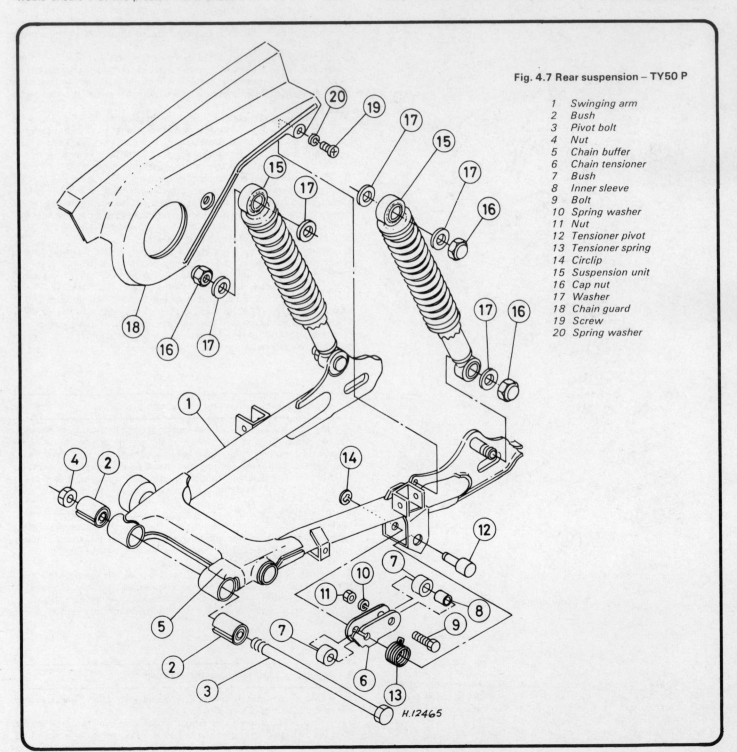

Fig. 4.7 Rear suspension – TY50 P

1   Swinging arm
2   Bush
3   Pivot bolt
4   Nut
5   Chain buffer
6   Chain tensioner
7   Bush
8   Inner sleeve
9   Bolt
10 Spring washer
11 Nut
12 Tensioner pivot
13 Tensioner spring
14 Circlip
15 Suspension unit
16 Cap nut
17 Washer
18 Chain guard
19 Screw
20 Spring washer

H.12465

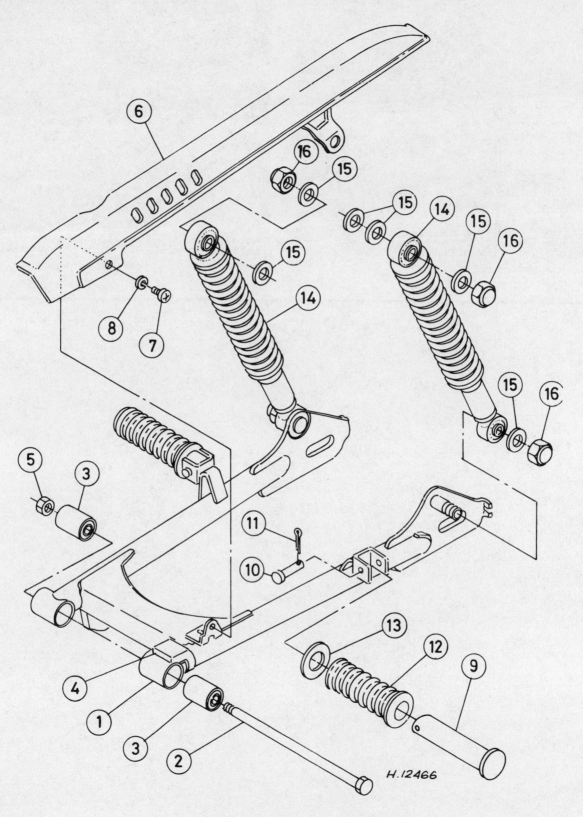

Fig. 4.8 Rear suspension – TY50 M

1	Swinging arm	5	Nut	9	Pillion footrest	13	Washer
2	Pivot bolt	6	Chain guard	10	Clevis pin	14	Suspension unit
3	Bush	7	Screw	11	Split pin	15	Washer
4	Chain buffer	8	Spring washer	12	Footrest rubber	16	Cap nut

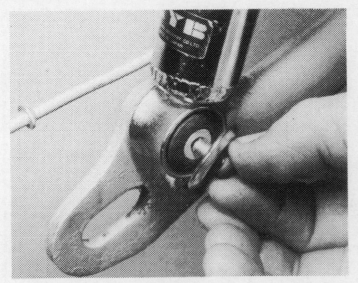
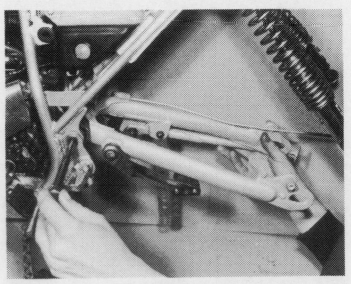

8.2 Remove suspension unit bottom mountings ...

8.3 ... and withdraw pivot bolt to release swinging arm

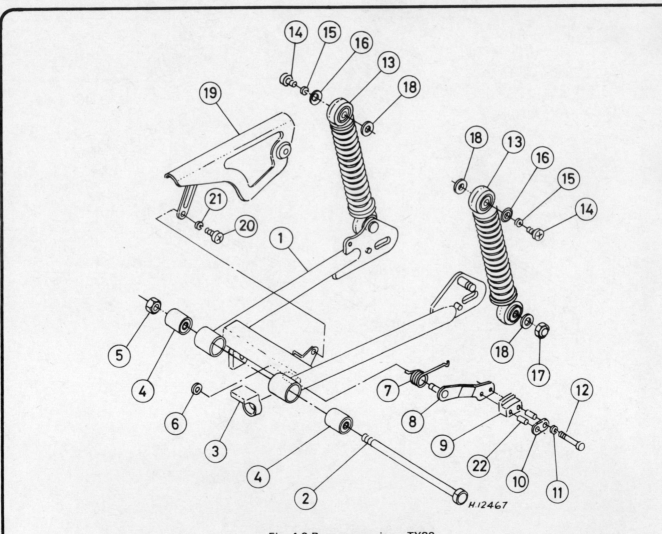

Fig. 4.9 Rear suspension – TY80

1	Swinging arm	7	Spring	13	Suspension unit	18 Washer
2	Pivot bolt	8	Chain tensioner	14	Screw	19 Chain guard
3	Chain buffer	9	Tensioner block	15	Spring washer	20 Screw
4	Bush	10	Retaining plate	16	Damping rubber	21 Spring washer
5	Nut	11	Spring washer	17	Cap nut	22 Spacer
6	Washer	12	Screw			

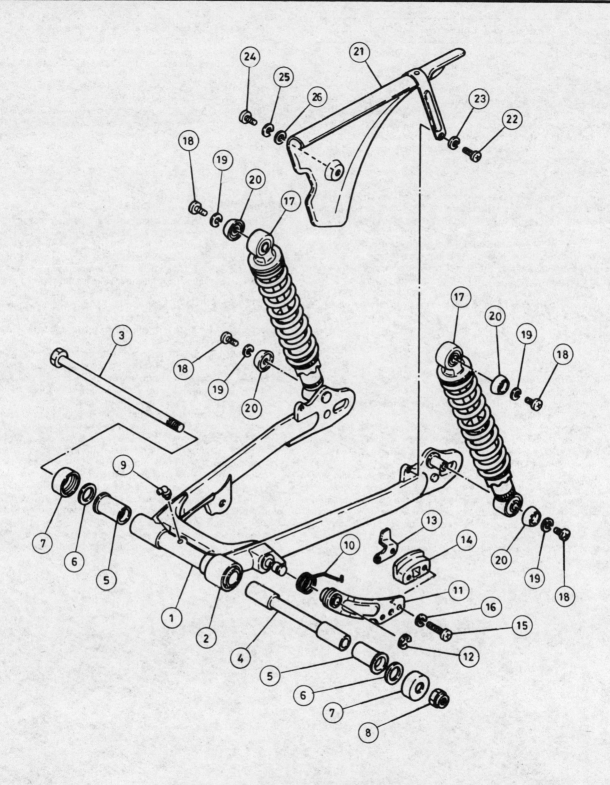

Fig. 4.10 Rear suspension – TY125 and 175

1	Swinging arm	8	Nut	15	Screw	21	Chain case
2	Chain guide	9	Grease nipple	16	Spring washer	22	Screw
3	Pivot bolt	10	Tensioner spring	17	Rear shock absorber	23	Spring washer
4	Inner sleeve	11	Tensioner arm	18	Screw	24	Screw
5	Outer bush	12	Circlip	19	Spring washer	25	Spring washer
6	Shim – as required	13	Chain guard	20	Washer	26	Washer
7	Dust seal	14	Tensioner block				

9 Swinging arm: examination and renovation

1 Dismantle as far as possible all components, removing the sealing caps and pushing out the hardened metal inner sleeve (where fitted) from each bearing. Thoroughly clean all components, removing all traces of dirt, corrosion and old grease.

2 Inspect closely all components looking for obvious signs of wear such as heavy scoring, or for damage such as cracks or distortion due to accidental impact. Any obviously damaged or worn component must be renewed. If the painted finish has deteriorated it is worth taking the opportunity to repaint the affected area, ensuring that the surface is correctly prepared beforehand.

3 Check the pivot bolt for wear. If the shank of the bolt is seen to be stepped or badly scored, then it must be renewed. Remove all traces of corrosion and hardened grease from the bolt before checking it for straightness by rolling it on a flat surface, such as a sheet of plate glass; if the bolt is not perfectly straight it must be renewed. Check also that its threads and those of the retaining nut are in good condition; the nut must be renewed if it has lost its self-locking quality.

4 Detach the nylon buffer from the left-hand pivot lug and inspect it for signs of excessive wear. If the buffer no longer protects the pivot lug from the final drive chain, it must be renewed.

5 Since the machines are fitted with pivot bearings of a very different type, they are described in separate sub-sections. Follow the instructions relevant to the machine being worked on.

TY50 and TY80 models

6 The bonded rubber bushes are an interference fit in the swinging arm mounting bosses, and are unlikely to wear or deteriorate until a high mileage has been covered, but wear may occur between the bush inner sleeves and the pivot, especially if the retaining nut has come loose. In normal service there should be no relative movement between the inner bushes and shaft; the fork movement is allowed by flexing of the bonded rubber.

7 Inspect the bush rubber for damage or separation from both inner and outer sleeves. Check also that the inner sleeves have not been moving on the shaft. If damage is evident, the bushes must be renewed. Driving the bushes from position is unlikely to prove successful, particularly if they have been in position for a long time and corrosion has taken place. Removal is accomplished most easily using a fabricated puller as shown in the accompanying illustration. The puller sleeve should have an internal diameter slightly greater than the outside diameter of the bush outer sleeve. If possible, use a high tensile nut and bolt because these will be better able to take the strain during use. New bushes may be drawn into place using the same puller. If difficulty is encountered in removing the old bushes it is recommended that the swinging arm be returned to a Yamaha Service Agent.

8 After fitting new bushes, the swinging arm fork may be refitted to the machine by reversing the dismantling procedure. No lubricant should be used on the bushes because they are made of rubber. The shaft should be lubricated to prevent corrosion between the shaft, bush inner bearings and frame members. A lithium soap-based grease is recommended.

TY125 and TY175 models

9 The pivot bearing consists of outer bushes pressed into each side of the swinging arm pivot lug, bearing on an inner hardened metal sleeve which is clamped in place by the pivot bolt. Any wear is usually found between inner sleeve and outer bushes and is checked, after thorough cleaning, by refitting the inner sleeve and feeling for free play. If any is found, or if serious wear marks such as scoring can be seen, the worn components must be renewed.

10 Remove the outer bushes only if renewal is required, since the method of removal will almost certainly damage the brittle material. If necessary, they can be displaced by inserting a drift from the end opposite to the bush shoulder and tapping with a hammer.

11 On refitting the bushes, clean the pivot lug thoroughly, removing any burrs or traces of corrosion, then coat both surfaces with grease. Assemble a drawbolt arrangement as shown in the accompanying illustration and carefully draw the bush into position, being careful to keep it absolutely square to its housing at all times.

12 Check the condition of the sealing caps, renewing any that are worn or damaged. The caps must be in good condition to prevent the entry of dirt or water.

13 Note that shims are fitted underneath the sealing caps to eliminate endfloat along the axis of the pivot bolt. When the swinging arm is installed in the frame and the pivot bolt retaining nut is tightened to the specified torque setting, there should be no discernible free play when the swinging arm is pulled and pushed from side to side, but the arm should be free to move; add (or subtract) shims to achieve this.

14 On reassembly use a high-quality molybdenum disulphide based grease to pack each bearing. Smear grease over the inside and outside of the inner sleeve as it is refitted, and to the sealing lips of each sealing cap.

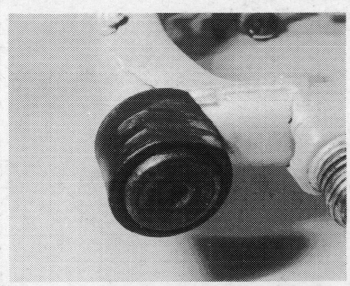

9.4 Renew nylon buffer if excessively worn from contact with chain

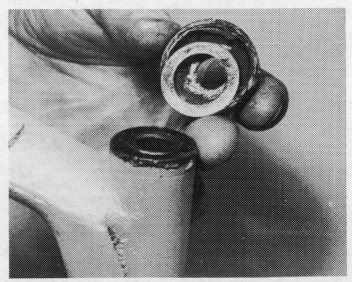

9.13 TY125 and 175 — fit shims as required behind sealing caps to eliminate swinging arm endfloat

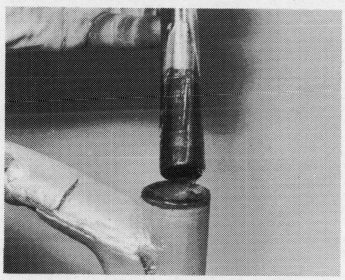

9.14 On reassemby pack pivot lug with grease, and smear grease over both surfaces of metal sleeve

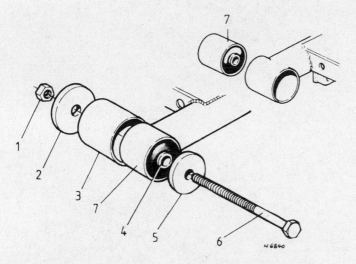

Fig. 4.11 Swinging arm bush removal tool – TY50 and 80

1	Nut	5	Thick washer
2	Thick washer	6	Bolt
3	Sleeve	7	Bush
4	Swinging arm		

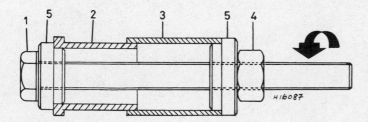

Fig. 4.12 Fitting the swinging arm pivot bushes – TY125 and 175

1	Drawbolt	4	Nut
2	Bush	5	Thick washer
3	Pivot lug		

10 Rear suspension units: removal, examination and refitting

1   The suspension units are withdrawn by removing the four retaining cap nuts or screws and plain washers, then pulling each unit sideways off its mounting lugs. Refitting is a straightforward reversal of the above, but care must be taken to refit correctly the plain washers. Tighten all four retaining nuts or screws to the specified torque setting.

2   The suspension units are sealed; if any wear or damage is found they must be renewed as a matched pair of complete units. Look for signs of oil leakage around the damper rod, worn mounting bushes, and any other signs of damage.

3   The spring preload setting of each suspension unit is adjustable to compensate for varying loads. The setting can be easily altered by using a C-spanner on the adjuster cam ring below the springs. Turning clockwise will increase the spring tension and stiffen the rear suspension, turning anti-clockwise will lessen the spring tension and therefore soften the ride.

11 Footrests, stands and controls: examination and renovation

1   At regular intervals all footrests, the stand and the brake pedal and gearchange lever should be checked and lubricated. Check that all mounting nuts and bolts are securely fastened, using the recommended torque wrench settings where these are given. Check that any securing split-pins are correctly fitted.

2   Check that the bearing surfaces at all pivot points are well greased and unworn, renewing any component that is excessively worn. If lubrication is required, dismantle the assembly to ensure that grease can be packed fully into the bearing surface. Return springs, where fitted, must be in good condition with no traces of fatigue and must be securely mounted.

3   If accident damage is to be repaired, check that the damaged component is not cracked or broken. Such damage may be repaired by welding, if the pieces are taken to an expert, but since this will destroy the finish, renewal is usually the most satisfactory course of action. If a component is merely bent it can be straightened after the affected area has been heated to a full cherry red, using a blowlamp or welding torch. Again the finish will be destroyed, but painted surfaces can be repainted easily, while chromed or plated surfaces can only be replated, if the cost is justified.

12 Speedometer head: removal, examination and refitting

1   This instrument must be carefully handled at all times and must never be dropped or held upside down. Dirt, oil, grease and water all have an equally adverse effect on it and so a clean working area must be provided if it is to be removed.

2   The instrument head is very delicate and should not be dismantled at home. In the event of a fault developing, the instrument should be entrusted to a specialist repairer or a new unit fitted. If a replacement unit is required, it is well worth trying to obtain a good secondhand item from a motorcycle breaker in view of the high cost of a new instrument.

3   Remember that a speedometer in correct working order is a statutory requirement in the UK. Apart from this legal necessity, reference to the odometer readings is the most satisfactory means of keeping pace with the maintenance schedules.

4   The instrument is removed by disconnecting the drive cable and unscrewing the mounting nuts or bolts. Where applicable, unplug the bulbholders to release the instrument. Reverse to refit the instrument.

13 Speedometer drive cable: examination and renovation

1   It is advisable to detach the speedometer drive cable from time to time in order to check whether it is adequately lubricated and whether the outer cable is compressed or damaged at any point along its run. A jerky or sluggish movement at the instrument head can often be attributed to a cable fault.

2   To grease the cable, uncouple both ends and withdraw the inner cable, where possible. After removing any old grease, clean the inner cable with a petrol soaked rag and examine the cable for broken strands or other damage. Do not check the cable for broken strands by passing it through the fingers or palm of the hand, this may well cause a painful injury if a broken strand snags the skin. It is best to wrap a piece of rag around the cable and pull the cable through it, any broken strands will snag the rag.

3   Regrease the cable with high melting point grease, taking care not to grease the last six inches closest to the instrument head. If this precaution is not observed, grease will work into the instrument and immobilise the sensitive movement.

4   The cable on all models is secured at its upper end by a large knurled ring which must be tightened or slackened using a suitable pair of pliers. Do not overtighten the knurled ring, or it will crack necessitating renewal of the complete cable. The lower end is pressed into a housing in the front brake backplate and secured by a wire circlip. Check that the sealing O-ring is in good condition.

5   When refitting a drive cable, always ensure that it has a smooth easy run to minimise wear, and check that it is secured where necessary by any clamps or ties provided for the purpose of keeping it away from any hot or moving parts.

14.2 Remove circlip to release speedometer drive assembly from brake backplate

14 Speedometer drive: location and examination

1   The speedometer drive is located in the front brake backplate; it is therefore necessary to withdraw the front wheel from the machine in order to gain access to the drive components.

2   To dismantle the drive assembly, remove the retaining circlip from the centre of the backplate, then lift out the thrust washer, driveplate, drive gear and second thrust washer. Turn the backplate over and remove the worm gear supporting bush, which is screwed into the backplate and must be unscrewed using either a specially-fabricated peg spanner or a thin-nosed punch and hammer. Withdraw the worm gear with the washer around its upper end. The oil seal should be renewed whenever it is disturbed in this way.

3   Carefully clean and inspect all the component parts. Any wear or damage will be immediately obvious and should be rectified by the renewal of the part concerned. The most likely areas of wear are on the tangs of the driveplate or on the teeth of either the drive gear or the drive pinion. If wear is encountered on the teeth of either of the latter components, it is advisable to renew both together. Note also the condition of the large oil seal set in the backplate. If this seal shows any sign of damage or deterioration it must be renewed to prevent grease from the drive assembly working through to the brake linings.

4   Reassemble the speedometer drive assembly in the reverse order of dismantling. Lightly lubricate all components with a good quality high melting-point grease.

15 Pedals – general – TY50P only

1   The TY50P models were fitted with pedals which could be locked in place to double as footrests, the pivot shaft passing through the swinging arm. A sprocket on the left-hand pedal drives the rear wheel via a short chain and a freewheel assembly mounted on the rear hub.

2   To dismantle the assembly, remove its nut and washer and drive out the cotter pin securing the left-hand pedal to the cross-shaft. Pull the shaft out to the right, noting the arrangement of the components in the locking assembly. Disengage the sprocket from the chain and withdraw the left-hand pedal.

3   Clean all components and renew any that are worn or damaged. The pedal return spring should be renewed if settled to much less than 71 mm (2.8 in). The pedals themselves have the same threads as pedal cycle components; note that the left-hand pedal employs a left-hand thread. Apply grease to all components on reassembly.

4   Remove the rear wheel to gain access to the freewheel. Remove the large circlip which secures the two split collets and unscrew the freewheel from its hub boss. Check that the freewheel rotates easily backwards but locks when rotated forwards; if faulty it must be renewed.

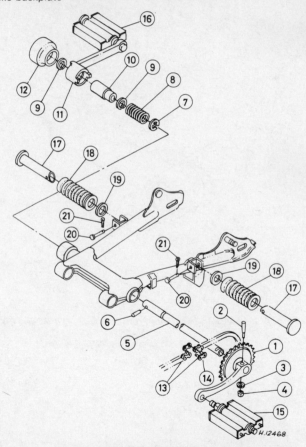

Fig. 4.13 Pedal gear – TY50 P

1 Left-hand pedal assembly	12 Rubber cover
2 Cotter pin	13 Chain
3 Spring washer	14 Connecting link
4 Nut	15 Left-hand pedal
5 Pedal shaft	16 Right-hand pedal
6 Dowel pin	17 Pillion footrest
7 Circlip	18 Footrest rubber
8 Pedal return spring	19 Washer
9 Circlip	20 Clevis pin
10 Locking bar	21 Split pin
11 Right-hand pedal crank	

Chapter 5 Wheels, brakes and tyres

Contents

Specifications

Wheels

	TY50 P, early TY50 M	late TY50 M	TY80	TY125, TY175
Rim size:				
Front	1.40 x 18	1.60 x 19	1.40 x 16	1.40 x 21
Rear	1.60 x 16	1.60 x 17	1.40 x 14	1.85 x 18
Rim maximum runout – radial and lateral	2.0 mm (0.08 in)			

Brakes

Front brake shoe diameter:	
TY50, TY125, TY175	110 mm (4.33 in)
Service limit	105 mm (4.13 in)
TY80	95 mm (3.74 in)
Service limit	90 mm (3.54 in)
Rear brake shoe diameter:	
TY50, TY80	110 mm (4.33 in)
Service limit	105 mm (4.13 in)
TY125, TY175	130 mm (5.12 in)
Service limit	125 mm (4.92 in)
Friction material thickness	4 mm (0.16 in)
Service limit	2 mm (0.08 in)
Return spring free length	34.5 mm (1.36 in)

Tyres

	TY50 P, early TY50 M	Late TY50 M	TY80	TY125, TY175
Size:				
Front	2.50 x 18-4PR	2.50 x 19-4PR	2.50 x 16-4PR	2.75 x 21-4PR
Rear	3.00 x 16-4PR	3.00 x 17-4PR	3.00 x 14-4PR	4.00 x 18-4PR
Recommended pressure:	TY50	TY80	TY125	TY175
Front	21 psi (1.5 kg/cm^2)	20 psi (1.4 kg/cm^2)	26 psi (1.8 kg/cm^2)	13 psi (0.9 kg/cm^2)
Rear	29 psi (2.0 kg/cm^2)	29 psi (2.0 kg/cm^2)	29 psi (2.0 kg/cm^2)	16 psi (1.1 kg/cm^2)

Torque wrench settings

Component	kgf m	lbf ft
Front wheel spindle nut:		
TY50, TY80	3.0 – 4.8	22.0 – 34.5
TY125, TY175	5.0 – 7.0	36.0 – 50.5
Front wheel spindle pinch bolt – TY50 P, early TY50 M only ..	1.4 – 2.2	10.0 – 16.0
Front wheel spindle cap nuts – TY125, TY175 only	1.0	7.0
Front brake torque arm retaining nuts – TY125, TY175 only ..	1.5	11.0
Rear wheel spindle nut:		
TY50	5.0 – 7.0	36.0 – 50.5
TY80	4.0 – 4.5	29.0 – 32.5
TY125, TY175	7.0 – 10.0	50.5 – 72.0
Sprocket retaining bolts:		
TY50, TY80	1.7 – 2.2	12.0 – 16.0
TY125, TY175	1.1 – 1.8	8.0 – 13.0
Rear brake torque arm/backplate retaining nut	3.0	22.0

1 General description

Tyres of conventional tubed type are fitted on to alloy or chromed steel rims that are laced to light alloy full-width hubs.

The brakes are of the single-leading shoe drum type, the front being cable-operated and the rear rod-operated.

2 Front wheel: removal and refitting

TY50 and TY80

1 Remove the split-pin (where fitted) and unscrew the spindle retaining nut. Lift the machine on to a strong wooden box or similar support so that it is supported securely with the front wheel clear of the ground. Disconnect the front brake cable and, if fitted, the speedometer cable. On TY50 P and early TY50 M models slacken fully the pinch bolt in the right-hand fork lower leg. Tap out the wheel spindle and withdraw the wheel, noting the arrangement of spacers and washers or sealing caps.

2 On reassembly, ensure that the tabs of the speedometer drive (where applicable) engage with the slots in the hub as the brake backplate is refitted, then check that all spacers are correctly installed. Offer up the wheel, engaging the slot in the backplate with the lug on the left-hand fork lower leg. Smear grease over the spindle and refit it. Refit and tighten by hand only the spindle nut then connect the front brake cable. Push the machine off its stand, apply the front brake hard and pump the forks up and down to settle them on the spindle. Where applicable tighten the pinch bolt to a torque setting of 1.4 – 2.2 kgf m (10.0 – 16.0 lbf ft).

3 With the brake applied hard to centralise the shoes and backplate on the drum, tighten the spindle nut to a torque setting of 3.0 – 4.8 kgf m (22.0 – 34.5 lbf ft). Fit a new split pin to secure the nut, spreading its ends correctly, and connect the speedometer drive cable (where applicable). Adjust the front brake as described in Routine Maintenance and check that the wheel revolves freely and that the brake works correctly.

TY125 and TY175

4 Remove the split-pin and unscrew the spindle retaining nut. Lift the machine on to a strong wooden box or similar support so that it is supported securely with the front wheel clear of the ground. Disconnect the front brake cable and, if fitted, the speedometer cable, then remove its split pin and retaining nut and lift the torque arm off the brake backplate. Slacken the two clamp nuts at the bottom of the right-hand fork leg and withdraw the wheel spindle. Remove the wheel from the machine. noting the arrangement of spacers.

5 On refitting, ensure that the tabs of the speedometer drive engage with the slots in the hub as the brake backplate is refitted, then check that all spacers are correctly installed and smear grease over the spindle. Offer up the wheel and refit the spindle. Rotate the hub right-hand spacer so that its slit aligns with the join between the spindle clamp and fork lower leg. Refit the spindle washer and nut, followed by the torque arm mounting bolt, lock washers and nut. Tighten all nuts by hand only, then connect the front brake and speedometer cable. Push the machine off its stand and pump the forks up and down to settle them on the spindle.

6 Using a spanner, rotate the spindle until the spacer slit is aligned correctly and hold it while the spindle nut is tightened to a torque setting of 5.0 – 7.0 kgf m (36.0 – 50.5 lbf ft). Tighten the spindle clamp nuts evenly to a torque setting of 1.0 kgf m (7.0 lbf ft), and the torque arm retaining nut to a setting of 1.5 kgf m (11.0 lbf ft). Spreading their ends securely, fit new split pins to the spindle and torque arm retaining nuts.

7 Adjust the front brake as described in Routine Maintenance and check that the wheel revolves freely and that the brake operates freely.

8 If the brake is found to be spongy and ineffective, it will need to be centralised on the drum. Remove their split pins and slacken off the spindle and torque arm retaining nuts. Apply the front brake hard and retighten first the spindle nut, then the torque arm nut, to their correct respective torque settings. Refit the split pins and recheck the brake adjustment.

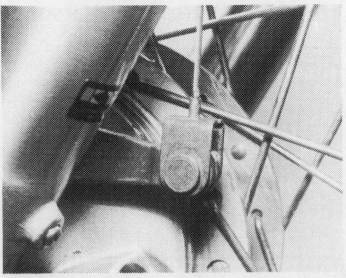

2.4 Disconnect brake cable at operating arm on brake backplate

2.5a Align tabs of speedometer drive with slots in wheel hub

2.5b Grease spindle and do not omit spacer and dust seal on refitting

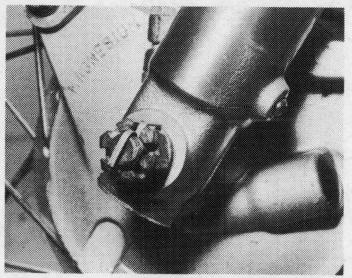

2.6a Tighten spindle nut to specified torque setting and secure with split pin ...

2.6b ... as shown, also secure brake torque arm retaining nuts

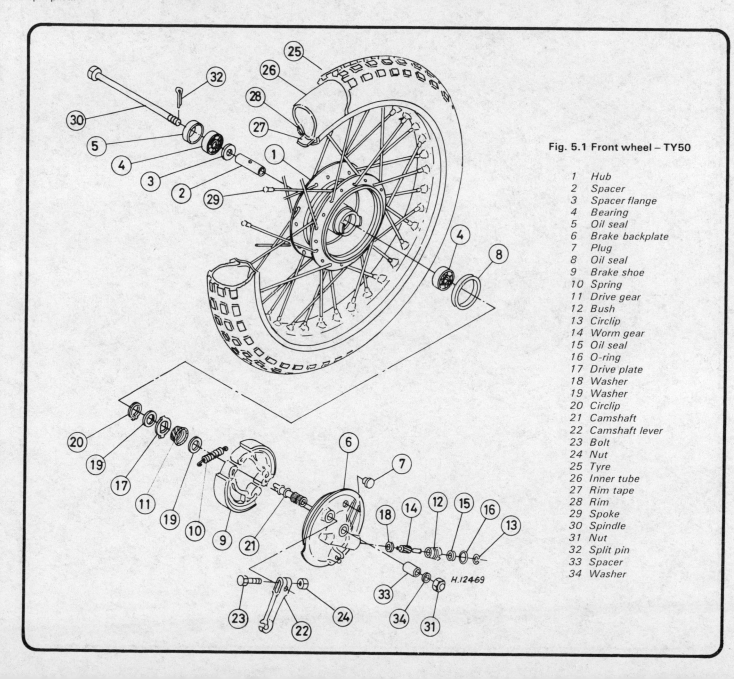

Fig. 5.1 Front wheel – TY50

1	Hub
2	Spacer
3	Spacer flange
4	Bearing
5	Oil seal
6	Brake backplate
7	Plug
8	Oil seal
9	Brake shoe
10	Spring
11	Drive gear
12	Bush
13	Circlip
14	Worm gear
15	Oil seal
16	O-ring
17	Drive plate
18	Washer
19	Washer
20	Circlip
21	Camshaft
22	Camshaft lever
23	Bolt
24	Nut
25	Tyre
26	Inner tube
27	Rim tape
28	Rim
29	Spoke
30	Spindle
31	Nut
32	Split pin
33	Spacer
34	Washer

H.12469

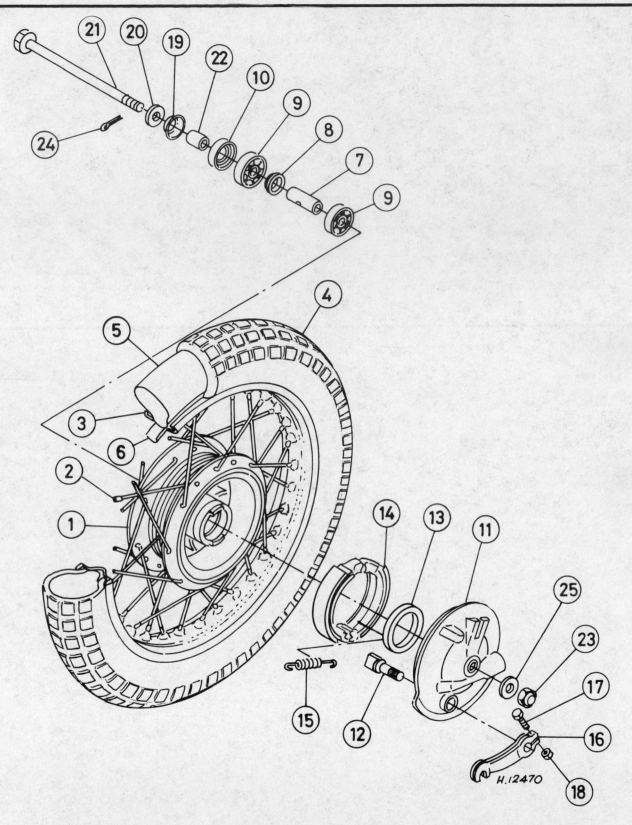

Fig. 5.2 Front wheel – TY80

1	Hub	8	Spacer flange	14	Brake shoe	20	Washer
2	Spoke	9	Bearing	15	Spring	21	Spindle
3	Rim	10	Oil seal	16	Camshaft lever	22	Spacer
4	Tyre	11	Brake backplate	17	Bolt	23	Nut
5	Inner tube	12	Camshaft	18	Nut	24	Split pin
6	Rim tape	13	Oil seal	19	Dust cover	25	Washer
7	Spacer						

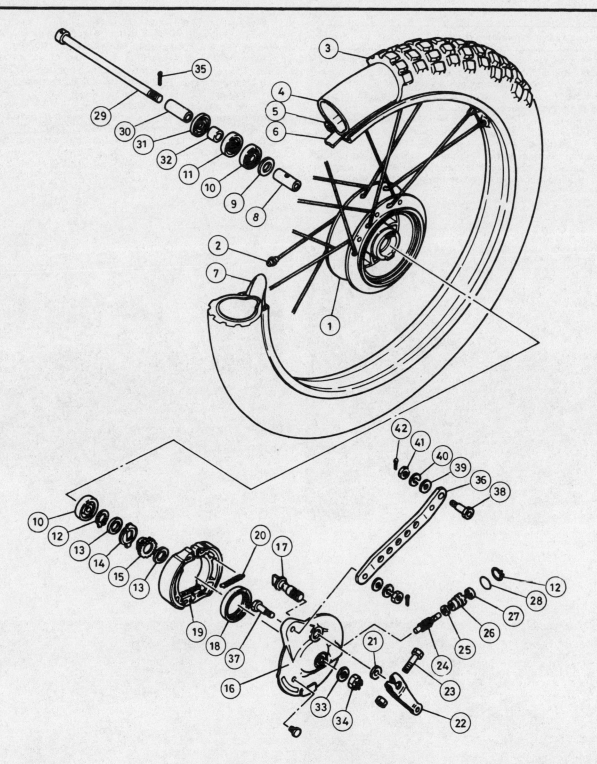

Fig. 5.3 Front wheel – TY125 and 175

1	Hub	12	Circlip	23	Bolt	33	Washer
2	Spoke	13	Washer	24	Worm gear	34	Nut
3	Tyre	14	Drive plate	25	Washer	35	Split pin
4	Inner tube	15	Drive gear	26	Bush	36	Torque arm
5	Rim	16	Brake backplate	27	Oil seal	37	Bolt
6	Rim tape	17	Camshaft	28	O-ring	38	Bolt
7	Security bolt	18	Oil seal	29	Spindle	39	Washer
8	Spacer	19	Brake shoe	30	Spacer	40	Spring washer
9	Spacer flange	20	Spring	31	Dust cover	41	Nut
10	Bearing	21	Seal	32	Spacer	42	Split pin
11	Oil seal	22	Camshaft lever				

3 Rear wheel: removal and refitting

1   Position a strong wooden box or blocks beneath the engine so that the machine is well supported with the rear wheel clear of the ground.

2   Pull the split-pin from the wheel spindle retaining nut and remove the nut. Remove the split-pin, nut and spring washer which retain the torque arm to the brake backplate and detach the arm.

3   Remove the nut from the brake rod end. Depress the brake pedal so the rod leaves the trunnion of the operating arm. Retain the nut, trunnion, spring and washer to prevent loss.

4   Lay a length of clean rag beneath the final drive chain and rotate the chain adjusters so that the wheel can be pushed forward and the chain lifted off the sprocket to be looped over the swinging arm fork end.

5   Support the wheel and pull out the spindle. If necessary, use a soft-metal drift and hammer to tap it from position . Remove the wheel, noting the fitted position of washers and spacers.

6   Refer to the accompanying figure and check all spacers are correctly fitted before lifting the wheel into position and engaging the chain on the sprocket. Grease the spindle before insertion and tap it lightly to seat it. Fit the retaining nut, finger-tight.

7   Reconnect the brake rod and torque arm. Renew the arm retaining washer if flattened, tighten the nut to the specified torque setting and fit a new split pin, where applicable.

8   Check the chain, rear brake and stop lamp rear switch adjustments as described in Routine Maintenance. Do not forget to tighten the spindle nut to the specified torque setting and to secure it with a new split pin.

3.6a Grease spindle before refitting – do not omit spacer or chain adjuster ...

3.6b ... or hub left-hand side dust seal/spacer

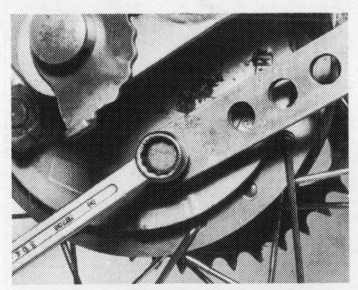

3.7a After chain and rear brake adjustment tighten securely torque arm mounting ...

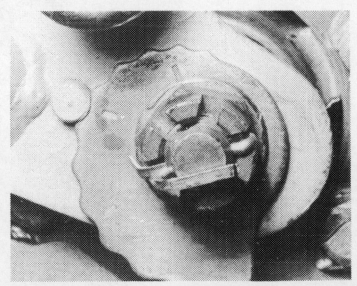

3.7b ... and wheel spindle nut

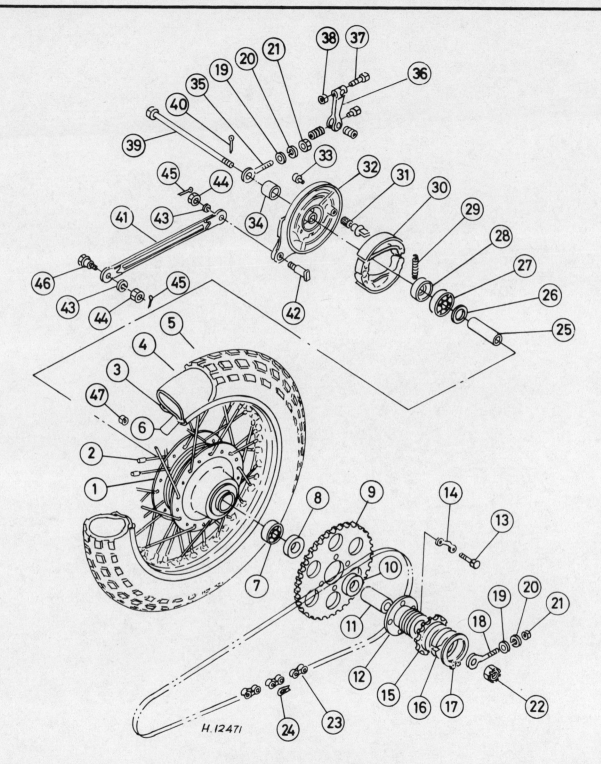

Fig. 5.4 Rear wheel – TY50

1	Hub	13	Bolt	25	Spacer	37	Bolt
2	Spoke	14	Lock washer	26	Spacer flange	38	Nut
3	Rim	15	Freewheel – TY50 P	27	Bearing	39	Spindle
4	Inner tube	16	Split collets – TY50 P	28	Oil seal	40	Split pin
5	Tyre	17	Circlip – TY50 P	29	Spring	41	Torque arm
6	Rim tape	18	Left-hand chain adjuster	30	Brake shoe	42	Bolt
7	Bearing	19	Washer	31	Camshaft	43	Spring washer
8	Oil seal	20	Spring washer	32	Brake backplate	44	Nut
9	Sprocket	21	Nut	33	Plug	45	Split pin
10	Oil seal	22	Nut	34	Spacer	46	Bolt
11	Spacer	23	Final drive chain	35	Right-hand chain adjuster	47	Nut
12	Freewheel boss – TY50 P	24	Chain joining link	36	Camshaft lever		

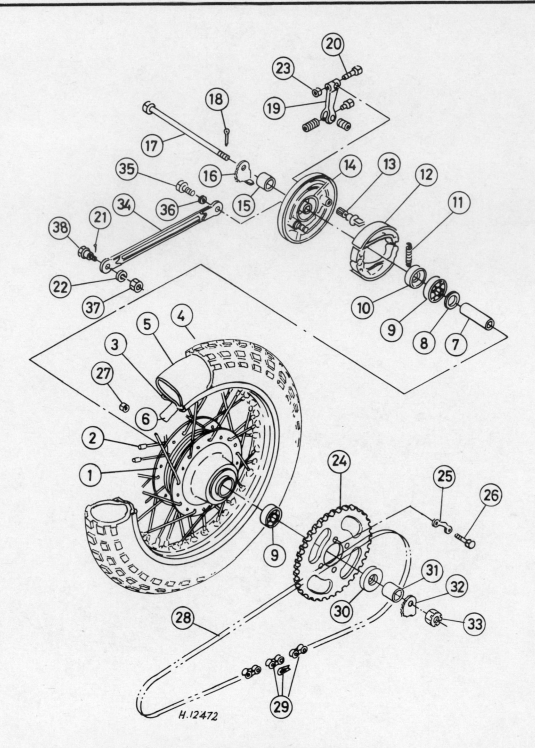

Fig. 5.5 Rear wheel — TY80

1 Hub	11 Spring	21 Split pin	30 Oil seal
2 Spoke	12 Brake shoe	22 Spring washer	31 Spacer
3 Rim	13 Camshaft	23 Nut	32 Left-hand snail cam
4 Tyre	14 Brake backplate	24 Sprocket	33 Nut
5 Inner tube	15 Spacer	25 Lock washer	34 Torque arm
6 Rim tape	16 Right-hand snail cam	26 Bolt	35 Bolt
7 Spacer	17 Spindle	27 Nut	36 Spring washer
8 Spacer flange	18 Split pin	28 Final drive chain	37 Nut
9 Bearing	19 Camshaft lever	29 Chain joining link	38 Bolt
10 Oil seal	20 Bolt		

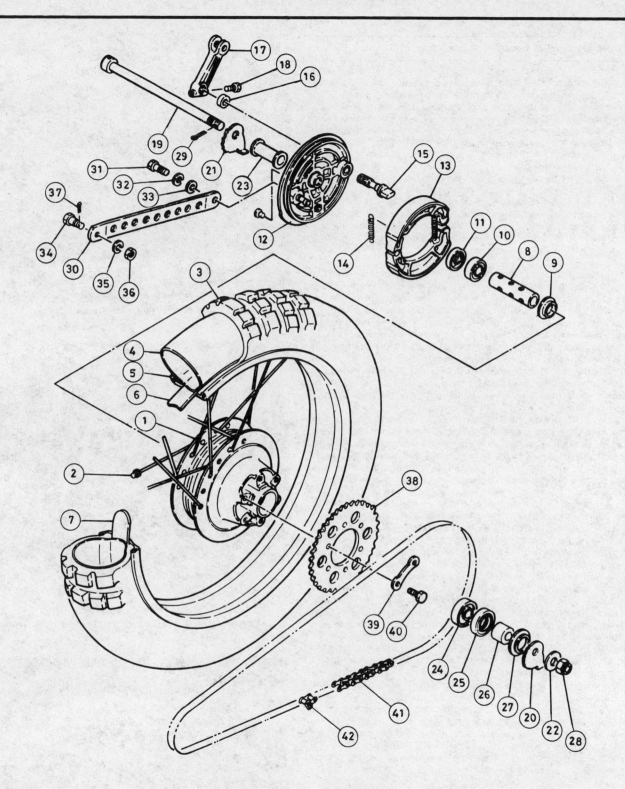

Fig. 5.6 Rear wheel – TY125 and 175

1 Hub	12 Brake backplate	23 Spacer	33 Washer
2 Spoke	13 Brake shoe	24 Bearing	34 Bolt
3 Tyre	14 Spring	25 Oil seal	35 Spring washer
4 Inner tube	15 Camshaft	26 Spacer	36 Nut
5 Rim	16 Seal	27 Dust cover	37 Split pin
6 Rim tape	17 Camshaft lever	28 Nut	38 Sprocket
7 Security bolt	18 Bolt	29 Split pin	39 Lock washer
8 Spacer	19 Spindle	30 Torque arm	40 Bolt
9 Spacer flange	20 Left-hand snail cam	31 Bolt	41 Chain
10 Bearing	21 Right-hand snail cam	32 Spring washer	42 Connecting link
11 Oil seal	22 Washer		

4 Wheel bearings: removal, examination and refitting

1   Remove the wheel from the machine as described in the previous Section of this Chapter, lift away the brake backplate and withdraw the spacer and sealing cap (if fitted) from the opposite side of the hub. On TY50 P models, flatten back the tab washers, unscrew the sprocket retaining bolts and lift away the freewheel hub and the sprocket.

2   Position the wheel on a work surface with its hub well supported by wooden blocks so that enough clearance is left beneath the wheel to drive out the left-hand bearing. Ensure the blocks are placed as close to the bearing as possible, to lessen the risk of distortion of the hub casting whilst the bearings are being removed or refitted.

3   Place the end of a long-handled drift against the upper face of the left-hand bearing and tap the bearing downwards out of the wheel hub. The spacer located between the two bearings may be moved sideways slightly in order to allow the drift to be positioned against the face of the bearing. Move the drift around the face of the bearing whilst drifting it out of position, so that the bearing leaves the hub squarely.

4   With the one bearing removed, the wheel may be lifted and the spacer withdrawn from the hub. Invert the wheel and remove the second bearing, using a similar procedure. The dust seal which fits against the right-hand bearing will be driven out as the bearing is removed. This seal should be closely inspected for any indication of damage, hardening or perishing and renewed if necessary. It is advisable to renew this seal as a matter of course if the bearings are found to be defective.

5   Remove all the old grease from the hub and bearings, giving the latter a final wash in petrol. Check the bearings for signs of play or roughness when they are turned. If there is any doubt about the condition of a bearing, it should be renewed.

6   If the original bearings are to be refitted, they should be repacked with the recommended grease before being fitted into the hub. New bearings must also be packed with the recommended grease. Ensure that the bearing recesses in the hub are clean and both bearings and recess mating surfaces lightly greased to aid fitting. Check the condition of the hub recesses for evidence of abnormal wear which may have been caused by the outer race of a bearing spinning. If evidence of this is found, and the bearing is a loose fit in the hub, it is best to seek advice from a Yamaha Service Agent or a competent motorcycle engineer. Alternatively, a proprietary product such as Loctite Bearing Fit may be used to retain the bearing outer race; this will mean, however, that the bearing housing must be cleaned and degreased before the locking compound can be used.

7   With the wheel hub and bearing thus prepared, fit the bearings and central spacer as follows. With the hub again well supported by the wooden blocks, drift the first of the two bearings into position. Use a soft-faced hammer in conjunction with a socket or length of metal tube which has an overall diameter which is slightly less than that of the outer race of the bearing, but which does not bear at any point on the bearing sealed surface or inner race. Tap the bearing into place against the locating shoulder machined in the hub, remembering that the sealed surface of the bearing must always face outwards. With the first bearing in place, invert the wheel, insert the central spacer with its flange on the right and pack the hub centre no more than $\frac{2}{3}$ full with high-melting point grease. Fit the second bearing, using the same procedure. Take great care to ensure that each of the bearings enters its housing correctly, that is, square to the housing, otherwise the housing surface may be broached.

8   Use the same method to refit the seal against the right-hand bearing. Refit the wheel to the machine.

5 Wheel sprocket: examination and renewal

1   Renew the sprocket if its teeth are hooked or badly worn. It is bad practice to renew the sprocket on its own; both drive sprockets should be renewed, preferably with the chain. Running old and new parts together will result in rapid wear.

2   Remove the wheel and bend back the locking tab from each sprocket securing bolt. Remove the bolts, tab washers and sprocket. When fitting, do not rebend the locking tabs, renew the washers if necessary. Tighten the bolts evenly and in a diagonal sequence to the specified torque setting.

4.3 Use drift as shown to remove wheel bearings

4.4a Withdraw central spacer once first bearing is removed

4.4b Oil seal can be levered out or drifted out with bearing

4.7a Bearings and oil seals are refitted as shown ...

4.7b ... ensuring that bearing sealed surfaces face outwards ...

4.8 ... and that oil seals are not damaged or distorted

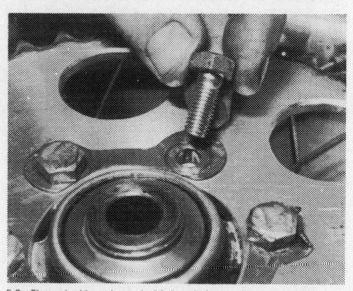

5.2a Flatten locking tabs and withdraw bolts to release sprocket

5.2b Secure bolts by bending up an unused part of lock washer as shown – renew lock washers when all tabs have been used

6 Brakes: examination and renovation

1   The brake assembly can be withdrawn from its hub after removal of the wheel from the machine.

2   Examine the condition of the brake linings. If they are worn beyond the specified limit the brake shoes should be renewed. The linings are bonded on and cannot be supplied separately.

3   If oil or grease from the wheel bearings has badly contaminated the linings, the brake shoes should be renewed. There is no satisfactory way of degreasing the lining material. Any surface dirt on the linings can be removed with a stiff-bristled brush. High spots on the linings should be carefully eased down with emery cloth.

4   Examine the drum surface for signs of scoring, wear beyond the service limit or oil contamination. All of these conditions will impair braking efficiency. Remove all traces of dust, preferably using a brass wire brush, taking care not to inhale any of it as it is of an asbestos nature, and consequently harmful. Remove oil or grease deposits using a petrol soaked rag.

5   If deep scoring is evident, due to the linings having worn through

to the shoe at some time, the drum must be skimmed on a lathe, or renewed. Whilst there are firms who will skim the drum without dismantling the wheel, it should be borne in mind that excessive skimming will change the radius of the drum in relation to the brake shoes, thereby reducing the friction area until extensive bedding in has taken place. Also full adjustment of the shoes may not be possible. If in doubt about this point, the advice of one of the specialist engineering firms who undertake this work should be sought.

6   It is a false economy to try to cut corners with brake components, the whole safety of both machine and rider being dependent on their good condition.

7   Removal of the brake shoes is accomplished by folding the shoes together so that they form a 'V'. With the spring tension relaxed, both shoes and springs may be removed from the brake backplate as an assembly. Detach the springs from the shoes and carefully inspect them for any signs of fatigue or failure. If in doubt, compare them with a new set of springs.

8   Before fitting the brake shoes, check that the brake operating cam is working smoothly and is not binding in its pivot. The cam can be removed by withdrawing the retaining bolt on the operating arm and pulling the arm off the shaft. Before removing the arm, it is advisable to mark its position in relation to the shaft, so that it can be relocated correctly, with the wear indicator pointer.

9   Remove any deposits of hardened grease or corrosion from the bearing surface of the brake cam and shoe by rubbing it lightly with a strip of fine emery paper or by applying solvent with a piece of rag. Lightly grease the length of the shaft and the face of the operating cam prior to reassembly. Clean and grease the pivot stub which is set in the backplate.

10 Check the condition of the O-ring which prevents the escape of grease from the end of the camshaft. If it is in any way damaged or perished, it must be renewed before the shaft is relocated in the backplate. Relocate the camshaft and align and fit the operating arm with the O-ring and wear indicator pointer. Tighten securely the pinch bolt.

11 Before refitting existing shoes, roughen the lining surface sufficiently to break the glaze which will have formed in use. Glasspaper or emery cloth is ideal for this purpose but take care not to inhale any of the asbestos dust that may come from the lining surface.

12 Fitting the brake shoes and springs to the brake backplate is a reversal of the removal procedure. Some patience will be needed to align the assembly with the pivot and operating cam whilst still retaining the springs in position; once the brake shoes are correctly aligned, they can be pushed back into position by pressing downwards to snap them into position. Do not use excessive force, as there is risk of distorting them permanently.

7 Tyres: removal, repair and refitting

1   To remove the tyre from either wheel, first detach the wheel from the machine. Deflate the tyre by removing the valve core, and when the tyre is fully deflated, push the bead away from the wheel rim on both sides so that the bead enters the centre well of the rim. Remove the locking ring and push the tyre valve into the tyre itself, then remove its nut and washer and press in the security bolt also.

2   Insert a tyre lever close to the valve and lever the edge of the tyre over the outside of the rim. Very little force should be necessary; if resistance is encountered it is probably due to the fact that the tyre beads have not entered the well of the rim, all the way round. If aluminium rims are fitted, damage to the soft alloy by tyre levers can be prevented by the use of plastic rim protectors.

3   Once the tyre has been edged over the wheel rim, it is easy to work round the wheel rim, so that the tyre is completely free from one side. Remove the inner tube and security bolt(s).

4   Now working from the other side of the wheel, ease the other edge of the tyre over the outside of the wheel rim that is furthest away. Continue to work around the rim until the tyre is completely free from the rim.

5   If a puncture has necessitated the removal of the tyre, reinflate the inner tube and immerse it in a bowl of water to trace the source of the leak. Mark the position of the leak, and deflate the tube. Dry the tube, and clean the area around the puncture with a petrol soaked rag. When the surface has dried, apply rubber solution and allow this to dry before removing the backing from the patch, and applying the patch to the surface.

6   It is best to use a patch of self vulcanising type, which will form a permanent repair. Note that it may be necessary to remove a protective covering from the top surface of the patch after it has sealed into position. Inner tubes made from a special synthetic rubber may require a special type of patch and adhesive, if a satisfactory bond is to be achieved.

7   Before replacing the tyre, check the inside to make sure that the article that caused the puncture is not still trapped inside the tyre. Check the outside of the tyre, particularly the tread area to make sure nothing is trapped that may cause a further puncture.

8   If the inner tube has been patched on a number of past occasions, or if there is a tear or large hole, it is preferable to discard it and fit a replacement. Sudden deflation may cause an accident, particularly if it occurs with the rear wheel.

9   To replace the tyre, inflate the inner tube sufficiently for it just to assume a circular shape, and then push the tube into the tyre so that it is enclosed completely. Lay the tyre on the wheel at an angle, and insert the valve through the rim tape and the hole in the wheel rim. Attach the locking ring on the first few threads, sufficient to hold the valve captive in its correct location.

10 Starting at the point furthest from the valve, push the tyre bead over the edge of the wheel rim until it is located in the central well. Continue to work around the tyre in this fashion until the whole of one side of the tyre is on the rim. It may be necessary to use a tyre lever during the final stages. Insert the security bolt(s) with their washers and nuts.

11 Make sure there is no pull on the tyre valve and again commencing with the area furthest from the valve, ease the other bead of the tyre over the edge of the rim. Finish with the area close to the valve, pushing the valve up into the tyre until the locking ring touches the rim. This will ensure that the inner tube is not trapped when the last section of bead is edged over the rim with a tyre lever.

12 Check that the inner tube is not trapped at any point. Reinflate the inner tube, and check that the tyre is seating correctly around the wheel rim. There should be a thin rib moulded around the wall of the tyre on both sides, which should be an equal distance from the wheel rim at all points. If the tyre is unevenly located on the rim, try bouncing the wheel when the tyre is at the recommended pressure. It is probable that one of the beads has not pulled clear of the centre well. When the tyre is correctly fitted and fully inflated tighten down hard the security bolt retaining nuts.

13 Always run the tyres at the recommended pressures and never under or over inflate. The correct pressures are given in the Specifications Section of this Chapter.

14 Tyre replacement is aided by dusting the side walls, particularly in the vicinity of the beads, with a liberal coating of french chalk. Washing up liquid can also be used to good effect, but this has the

6.7 Removing brake shoes – apply same method on refitting

Tyre changing sequence - tubed tyres

 A Deflate tyre. After pushing tyre beads away from rim flanges push tyre bead into well of rim at point opposite valve. Insert tyre lever adjacent to valve and work bead over edge of rim.

Use two levers to work bead over edge of rim. Note use of rim protectors **B**

C Remove inner tube from tyre

When first bead is clear, remove tyre as shown **D**

 E When fitting, partially inflate inner tube and insert in tyre

Work first bead over rim and feed valve through hole in rim. Partially screw on retaining nut to hold valve in place. **F**

 G Check that inner tube is positioned correctly and work second bead over rim using tyre levers. Start at a point opposite valve.

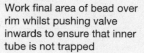

Work final area of bead over rim whilst pushing valve inwards to ensure that inner tube is not trapped **H**

disadvantage, where steel rims are used, of causing the inner surface of the wheel rim to rust.

15 Never replace the inner tube and tyre without the rim tape in position. If this precaution is overlooked there is a good chance of the ends of the spoke nipples chafing the inner tube and causing a crop of punctures.

16 Never fit a tyre that has a damaged tread or sidewalls. Apart from legal aspects, there is a very great risk of a blowout, which can have very serious consequences on a two wheeled vehicle.

8 Valve cores and caps: general

1 Valve cores seldom give trouble, but do not last indefinitely. Dirt under the seating will cause a puzzling 'slow-puncture'. Check that they are not leaking by applying spittle to the end of the valve and watching for air bubbles.

2 A valve cap is a safety device, and should always be fitted. Apart from keeping dirt out of the valve, it provides a second seal in case of valve failure, and may prevent an accident resulting from sudden deflation.

Chapter 6 Electrical system

Contents

Specifications

Battery

Make	Yuasa or FB
Type:	
TY50	6N4A-4D
TY125	6N4B-2A-3
Voltage	6 volt
Rating	4 Ah
Earth	Negative

Fuse rating .... 10 amp

Flywheel generator .... see main text

Bulbs

	TY50	TY125	TY175
Headlamp	7V, 18/18W	6V, 25/25W	6V, 25/25W
Stop/tail lamp	6V, 17/5.3W or 21/5W	6V, 21/5W	6V, 17/5.3W
Parking lamp	N/App	6V, 3W	N/App
Turn signal lamps	6V, 10W	6V, 18W	N/App
Instrument illuminating and warning lamps	6V, 3W	N/App	6V, 1.5W

1 General description

Except for TY80 models, which have only the basic ignition system described in Chapter 3, all models are fitted with an electrical system which is powered by the remaining coil of the flywheel generator asembly.

On TY50 and TY125 models part of the generator's output is diverted through the ignition and lighting switch to power the headlamp alone. The remainder of its output, in the form of alternating current, is converted to direct current by a half-wave silicon rectifier and fed to the battery which powers the stop lamp, horn, turn signals (where fitted), the tail lamp and any ancillary components fitted. The TY125 is fitted with an additional feed from the coil to provide an increased rate of charge at night.

TY175 models are fitted with an extremely simple system in which one feed from the coil (the yellow wire) provides the current for the stoplamp, and, if fitted, the horn, while the other feed (the yellow/red wire) provides the current for the lights. With no form of voltage control, this system relies heavily on being fitted with bulbs of the correct wattage and on the switches being in good condition.

2 Testing the electrical system

1 Simple continuity checks, for instance when testing switch units, wiring and connections, can be carried out using a battery and bulb arrangement to provide a test circuit. For most tests described in this chapter, however, a pocket multimeter should be considered essential. A basic multimeter capable of measuring volts and ohms can be bought for a very reasonable sum and will prove an invaluable tool. Note that separate volt and ohm meters may be used in place of the multimeter, provided those with the correct operating ranges are available. In addition, if the generator output is to be checked, an ammeter of 0-5 amperes range will be required.

2 Care must be taken when performing any electrical test, because some of the electrical components can be damaged if they are incorrectly connected or inadvertently shorted to earth. This is particularly so in the case of electronic components. Instructions regarding meter probe connections are given for each test, and these should be read carefully to preclude accidental damage occurring.

3 Where test equipment is not available, or the owner feels unsure of the procedure described, it is strongly recommended that pro

fessional assistance is sought. Errors made through carelessness or lack of experience can so easily lead to damage and need for expensive replacement parts.

4   A certain amount of preliminary dismantling will be necessary to gain access to the components to be tested. Normally, removal of the seat and side panels will be required, with the possible addition of the fuel tank and headlamp unit to expose the remaining components.

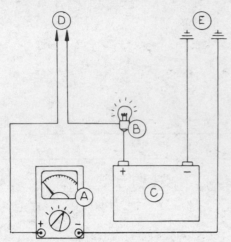

Fig. 6.1 Simple circuit testing arrangement

A	Multimeter	D	Positive probe
B	Bulb	E	Negative probe
C	Battery		

3   Wiring: layout and examination

1   The wiring harness is colour-coded and will correspond with the accompanying wiring diagram. When socket connections are used, they are designed so that reconnection can be made in the correct position only.

2   Visual inspection will usually show whether there are any breaks or frayed outer coverings which will give rise to short circuits. Occasionally a wire may become trapped between two components, breaking the inner core but leaving the more resilient outer cover intact. This can give rise to mysterious intermittent or total circuit failure. Another source of trouble may be the snap connectors and sockets, where the connector has not been pushed fully home in the outer housing, or where corrosion has occurred.

3   Intermittent short circuits can often be traced to a chafed wire that passes through or is close to a metal component such as a frame member. Avoid tight bends in the lead or situations where a lead can become trapped between casings.

4   Flywheel generator: checking the lighting circuit output

1   The output of the generator must be checked if it is suspected of being too high or too low. If too low the lights will be very dim, but if too high it will cause frequent bulb-blowing, indicated by the characteristic melted filaments. Be careful to check first for incorrect bulb wattages, poor connections or switches and for damaged wiring.

2   The output is measured by connecting the positive (+) terminal of an ac voltmeter to the terminal of the yellow wire from the flywheel generator at the point where it joins the main loom, adjacent to the oil tank. The meter negative (-) terminal is connected to a good earth point on the engine. On TY175 models, note that the test is repeated on the yellow/red wire.

3   Start the engine, switch the lights on (main beam) and note the readings obtained at the following engine speeds:

	TY50	TY125	TY175 – yellow wire	TY175 – yellow/red wire
2000 rpm	6.0V min	5.5V min	3.0V min	5.8V min
7000 rpm	N/Av	8.5V max	9.0V max	8.5V max

4   If the readings obtained are markedly different from those given, check again for faulty wiring, connections or switches, noting that similar results should be obtainable anywhere in the ac lighting circuit. If all is well, proceed to check the source coil as described in Section 6 of this Chapter.

5   On TY50 and TY125 models, if the output is too high and all other components have been found to be in good condition, check the charging system as described in the next Section. If the charging circuit voltage is low due to a faulty battery, rectifier or connections, the lighting circuit voltage will be raised.

5   Flywheel generator: checking the charging circuit output

1   On TY50 and TY125 models only, the output of the charging circuit must be checked if the battery is consistently over- or under-charged. Check first that the fault is not due to faulty wiring, connections, switches or to incorrect bulb wattages. Check also that the battery is capable of meeting the demand placed on it; if a heavy use is made of turn signals, stop lamp and horn, especially at night, it is possible that the battery cannot keep pace even though the system is functioning correctly.

2   Note that for the test results to be accurate, the battery must be in good condition and fully charged, as described elsewhere in this Chapter.

3   Disconnect the battery positive (+) wire at the connector joining it to the main loom, then connect the positive (+) terminal of a dc voltmeter to the terminal of the wire from the main loom and the meter negative (-) terminal to the battery wire terminal. Start the engine and note the readings obtained at the engine speeds given, testing once with the lights switched off for day time readings, then with the lights switched on for night time readings. A value of at least 6.5 volts should be obtained on TY50 models at 2000 rpm; for TY125 models results should be as follows:

	Day	Night
2000 rpm	8.5 volts	6.5 volts
7000 rpm	8.8 volts	7.0 volts

4   Substitute an ammeter for the voltmeter and repeat the test as described above. A value of at least 0.1 amp should be obtained on TY50 models at 2000 rpm; for TY125 models results should be as follows for day time readings:

2000 rpm	1.8 ± 0.5 amp minimum
7000 rpm	3.3 ± 0.5 amp maximum

As charging does not begin until 1500 rpm, results for TY125 models readings should be as follows:

4000 rpm	1.3 ± 0.3 amp minimum
7000 rpm	1.9 ± 0.3 amp maximum

5   If the readings obtained are markedly different from those given, check first the generator source coil, as described in the following Section. If this proves to be in good condition, check the rectifier, wiring and switches.

6   Flywheel generator: checking the coils

1   The condition of the generator source coil can be checked by measuring its resistance. Trace the generator lead from the crankcase top to the connectors joining it to the main loom, disconnect the wires to be tested, and measure the resistance between the wire terminal and a good earth point on the engine using a multimeter or ohmmeter set to the x 1 ohm scale. Note that two values are given for the TY50 models, the first relates to the TY50 P and early TY50 M models and the second to later TY50 M models from engine/frame no. 1G7-400101 (model code 2KO). Results should be as follows:

AC lighting coil

TY50 (early)	Yellow to earth	0.37 ohm ± 10% @ 20°C (68°F)
TY50 (later)	Yellow to earth	0.26 ohm ± 10% @ 20°C (68°F)
TY125	Yellow to earth	0.16 ohm ± 10% @ 20°C (68°F)
TY175	Yellow to earth	1.07 ohm ± 10% @ 20°C (68°F)
TY175	Yellow/red to earth	0.49 ohm ± 10% @ 20°C (68°F)

Charging coil

TY50 (early)	Green to earth	0.25 ohm ± 10% @ 20°C (68°F)
TY50 (later)	Green/red to earth	0.46 ohm ± 10% @ 20°C (68°F)
TY125-day	Green to earth	0.77 ohm ± 10% @ 20°C (68°F)
TY125-night	Green/red to earth	0.28 ohm ± 10% @ 20°C (68°F)

2   If any test produces results different from those given the coil is defective, but note that a reading of infinite resistance would indicate a broken wire (open circuit) and a reading of very little resistance would indicate a trapped wire (short circuit), either of which may be easily repaired by the owner. If the coil must be renewed, remember that an auto-electrical specialist may well be able to rewind it, saving some of the cost of a new item.

3   To renew the coil, the generator stator must be removed as described in Chapter 1 so that the coil connections can be unsoldered and the mounting screws removed to release the coil. Note carefully at which point the wires are connected and ensure that the new soldered joint is secure and well insulated on refitting; if in any doubt take the stator to an expert. Check carefully on refitting that the coil does not foul the rotor at any point, and that the air gap is the same between each coil pole and the rotor magnets.

7   Rectifier: location and testing

1   The silicon diode rectifier fitted to these machines is a small rectangular black plastic block with two male spade terminals projecting from its underside, which is retained by a single screw to the right-hand side of the frame top tubes, immediately behind the steering head. It requires no maintenance at all, save a periodic check that it is clean, and that both it and its connections are securely fastened.

2   The rectifier consists of a small diode and serves to convert the ac output of the flywheel generator into dc to charge the battery. It should be thought of as a one-way valve, in that it will allow the current to flow in one direction only, thus blocking half of the output wave from the generator.

3   Before removing the unit, identify the polarity of the two terminals by the colour of the wire leading to each one. The red wire leads to the positive (+) terminal and the white wire to the negative (-) or ac terminal.

4   Using a multimeter set to the resistance mode, check for continuity between the two terminals. There should only be continuity from the negative (-) to the positive (+) terminal. This direction of flow may be shown by an arrow on the surface of the unit. If there is continuity in the reverse direction, or if resistance is measured in both directions, the rectifier is faulty and must be renewed. No repair is possible.

8   Battery: examination and maintenance

1   The battery is housed in a tray located underneath the seat on TY50 models, and on TY125 models it is mounted in a carrier attached to the rear right-hand side of the frame.

2   The transparent plastic case of the battery permits the upper and lower levels of the electrolyte to be observed without disturbing the battery. Maintenance is normally limited to keeping the electrolyte level between the prescribed upper and lower limits and making sure that the vent tube is not blocked. The lead plates and their separators are also visible through the transparent case, a further guide to the general condition of the battery. If electrolyte level drops rapidly, suspect over-charging and check the system.

3   Unless acid is spilt, as may occur if the machine falls over, the electrolyte should always be topped up with distilled water to restore the correct level. If acid is spilt onto any part of the machine, it should be neutralised with an alkali such as washing soda or baking powder and washed away with plenty of water, otherwise serious corrosion will occur. Top up with sulphuric acid of the correct specific gravity (1.260 to 1.280) only when spillage has occurred. Check that the vent pipe is well clear of the frame or any of the other cycle parts.

4   It is seldom practicable to repair a cracked battery case because the acid present in the joint will prevent the formation of an effective seal. It is always best to renew a cracked battery, especially in view of the corrosion which will be caused if the acid continues to leak.

5   If the machine is not used for a period of time, it is advisable to remove the battery and give it a 'refresher' charge every six weeks or so from a battery charger. The battery will require recharging when the specific gravity falls below 1.260 (at 20°C -68°F). The hydrometer reading should be taken at the top of the meniscus with the hydrometer vertical. If the battery is left discharged for too long, the plates will sulphate. This is a grey deposit which will appear on the surface of the plates, and will inhibit recharging. If there is sediment on the bottom of the battery case, which touches the plates, the battery needs to be renewed. Prior to charging the battery refer to the following Section for correct charging rate and procedure. If charging from an external source with the battery on the machine, disconnect the leads, or the rectifier will be damaged.

6   Note that when moving or charging the battery, it is essential that that following basic safety precautions are taken:

(a)   Before charging check that the battery vent is clear or, where no vent is fitted, remove the combined vent/filler caps. If this precaution is not taken the gas pressure generated during charging may be sufficient to burst the battery case, with disastrous consequences.

(b)   Never expose a battery on charge to naked flames or sparks. The gas driven off by the battery is highly explosive.

(c)   If charging the battery in an enclosed area, ensure that the area is well ventilated.

(d)   Always take great care to protect yourself against accidental spillage of the sulphuric acid contained within the battery. Eyeshields should be worn at all times. If the eyes become contaminated with acid they must be flushed with fresh water immediately and examined by a doctor as soon as possible. Similar attention should be given to a spillage of acid on the skin.

Note also that although, should an emergency arise, it is possible to charge the battery at a more rapid rate than that stated in the following Section, this will shorten the life of the battery and should therefore be avoided if at all possible.

7   Occasionally, check the condition of the battery terminals to ensure that corrosion is not taking place, and that the electrical connections are tight. If corrosion has occurred, it should be cleaned away by scraping with a knife and then using emery cloth to remove the final traces. Remake the electrical connections whilst the joint is still clean, then smear the assembly with petroleum jelly (NOT grease) to prevent recurrence of the corrosion. Badly corroded connections can have a high electrical resistance and may give the impression of complete battery failure.

9   Battery: charging procedure

1   Whilst the machine is used on the road it is unlikely that the battery will require attention other than routine maintenance because the generator will keep it fully charged. However, if the machine is used for a succession of short journeys only, mainly during the hours of darkness when the lights are in full use, it is possible that the output from the generator may fail to keep pace with the heavy electrical demand, especially if the machine is parked with the lights switched on. Under these circumstances it will be necessary to remove the battery from time to time to have it charged independently.

2   The normal maximum charging rate for any battery is 1/10 the rated capacity. Hence the charging rate for the 4 Ah battery is 0.4 amp. A slightly higher charge rate may be used in emergencies only, but this should never exceed 1 amp.

3   Ensure that the battery/charger connections are properly made, ie the charger positive (usually coloured red) lead to the battery positive (the red wire) lead, and the charger negative (usually coloured black or blue) lead to the battery negative (the black/white wire) lead. Refer to the previous Section for precautions to be taken during charging. It is especially important that the battery cell cover plugs are removed to eliminate any possibility of pressure building up in the battery and cracking its casing. Switch off the charger if the cells become overheated, ie over 45°C (117°F).

4   Charging is complete when the specific gravity of the electrolyte rises to 1.260 – 1.280 at 20°C (68°F). A rough guide to this state is when all cells are gassing freely. At the normal (slow) rate of charge this will take between 3 – 15 hours, depending on the original state of charge of the battery.

5   If the higher rate of charge is used, never leave the battery charging for more than 1 hour as overheating and buckling of the plates will inevitably occur.

10 Fuse: location and renewal

1 The electrical system is protected by a single fuse of 10 amp rating. It is retained in a plastic casing set in the battery position (+) terminal lead, and is clipped to a holder behind the left-hand side panel. If the spare fuse is ever used, replace it with one of the correct rating as soon as possible.

2 Before renewing a fuse that has blown, check that no obvious short circuit has occurred, otherwise the replacement fuse will blow immediately it is inserted. It is always wise to check the electrical circuit thoroughly, to trace the fault and eliminate it.

3 When a fuse blows while the machine is running and no spare is available, a 'get you home' remedy is to remove the blown fuse and wrap it in silver paper before replacing it in the fuse holder. The silver paper will restore the electrical continuity by bridging the broken fuse wire. This expedient should never be used if there is evidence of short circuit or other major electrical faults, otherwise more serious damage will be caused. Replace the 'doctored' fuse at the earliest possible opportunity, to restore full circuit protection.

11 Switches: general

1 While the switches should give little trouble, they can be tested using a multimeter set to the resistance function or a battery and bulb test circuit. Using the information given in the wiring diagram at the end of this Manual, check that full continuity exists in all switch positions and between the relevant pairs of wires. When checking a particular circuit follow a logical sequence to eliminate the switch concerned.

2 As a simple precaution always disconnect the battery before removing any of the switches, to prevent the possibility of a short circuit. Most troubles are caused by dirty contacts, but in the event of the breakage of some internal part, it will be necessary to renew the complete switch.

3 If a switch is tested and found to be faulty, there is nothing to be lost by attempting a repair. It may be that worn contacts can be built up with solder, or that a broken wire terminal can be repaired, again using a soldering iron. The handlebar switches can all be dismantled to a greater or lesser extent. It is, however, up to the owner to decide if he has the skill to carry out this sort of work.

4 While none of the switches require routine maintenance, some regular attention will prolong their life. In the author's experience, the regular and constant application of WD40 or a similar water-dispersant spray not only prevents problems occurring due to water-logged switches and the resulting corrosion, but also makes the switches much easier and more positive to use. Alternatively, the switch may be packed with a silicone-based grease to achieve the same result.

12 Bulbs: renewal

1 All bulbs on these machines are of the bayonet type and can be released by pushing in, turning anti-clockwise and pulling from the holder. The tail/stop lamp is fitted with a double filament bulb which has offset pins to prevent unintentional reversal in its holder.

2 Gain access to the main headlamp bulb holder by detaching the rim, complete with reflector and glass, from the shell or nacelle. Unclip the holder from the centre of the reflector to expose the bulb. TY125 models have a parking lamp fitted; pull the holder from the reflector to expose the bulb.

3 The turn signal bulbs can be removed after detachment of the plastic lens, which will be secured by screws or clipped in position. Take care not to tear the lens seal (where fitted); this keeps moisture and dirt away from the backplate and electrical contacts and must be renewed if damaged.

4 The instrument and warning light bulbs are fitted in rubber holders which can be unplugged from the base of the instrument once it is exposed.

5 Clean any corrosion or moisture from the holder and check its contacts are free to move when depressed. Fit the bulb and lens. A lens can be cracked if its securing screws are overtightened. After tightening the headlamp rim, check beam alignment.

6 UK regulations stipulate that the headlamp must be aligned so that the light will not dazzle a person standing at a distance greater than 25 feet from the lamp, whose eye level is not less than 3 feet 6 inches above that plane. It is easy to approximate this setting by placing the machine 25 feet away from a wall, on a level road, and setting the dip beam height so that it is concentrated at the same height as the distance of the centre of the headlamp from the ground. The rider must be seated normally during this operation.

7 To effect beam vertical alignment, loosen the headlamp shell or nacelle securing screws and pivot the lamp up or down, as required.

13 Horn: location and testing

1 The horn is mounted on a flexible steel bracket to the bottom yoke. No maintenance is required other than regular cleaning to remove road dirt and occasional spraying with WD40 or a similar water dispersant lubricant to minimise internal corrosion.

2 Different types of horn may be fitted; if a screw and locknut is provided on the outside of the horn, the internal contacts may be adjusted to compensate for wear and to cure a weak or intermittent horn note. Slacken the locknut and rotate slowly the screw until the clearest and loudest note is obtained, then retighten the locknut. If no means of adjustment is provided on the horn fitted, it must be renewed.

3 If the horn fails to work, first check that the power is reaching it

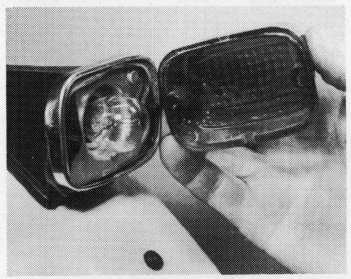

12.1 Except for headlamp, all bulbs are bayonet fitting — be careful not to crack lens when tightening screws

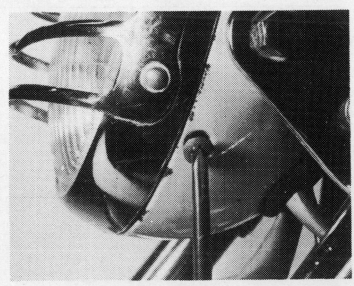

12.2a Remove rim retaining screw(s) ...

by disconnecting the wires. Substitute a 6 volt bulb, switch on the ignition and press the horn button. If the bulb lights, the circuit is proved good and the horn is at fault; if the bulb does not light, there is a fault in the circuit which must be found and rectified.

4   To test the horn itself, connect a fully-charged 6 volt battery directly to the horn. If it does not sound, a gentle tap on the outside may free the internal contacts. If this fails, the horn must be renewed as repairs are not possible.

---

14 Turn signal relay: location and testing

1   The relay is a cylindrical sealed metal unit rubber-mounted under the tank on TY125 models and under the seat on TY50 models.

2   If the turn signal lamps cease to function correctly, there may be any one of several possible faults responsible which should be checked before the relay is suspected. First check that the turn signal lamps are correctly mounted and that all the earth connections are clean and tight. Check that the bulbs are of the correct wattage and that corrosion has not developed on the bulbs or in their holders. Any such corrosion must be thoroughly cleaned off to ensure proper bulb contact. Also check that the turn signal switch is functioning correctly and that the wiring is in good order. Finally, ensure that the battery is fully charged.

3   Faults in any one or more of the above items will produce symptoms for which the turn signal relay may be blamed unfairly. If the fault persists even after the preliminary checks have been made, the relay must be at fault. Unfortunately the only practical method of testing the relay is to substitute a known good one.

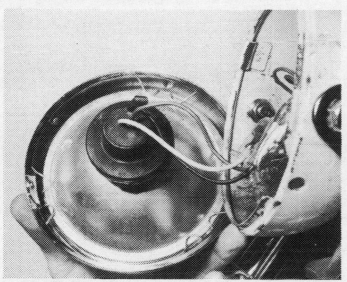

12.2b ... to release headlamp assembly

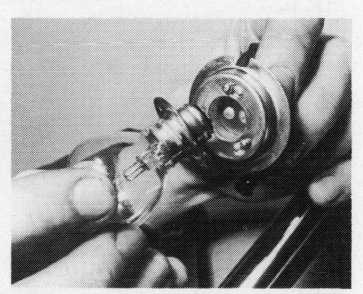

12.2c Align pins with dots in bulb flange on refitting

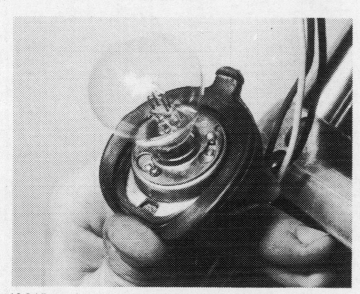

12.2d Ensure bulb holder is located correctly in mounting rubber

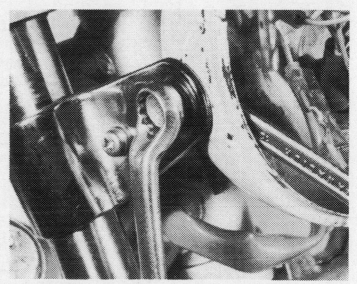

12.7 Slacken mounting bolts and tilt shell for headlamp beam alignment

154

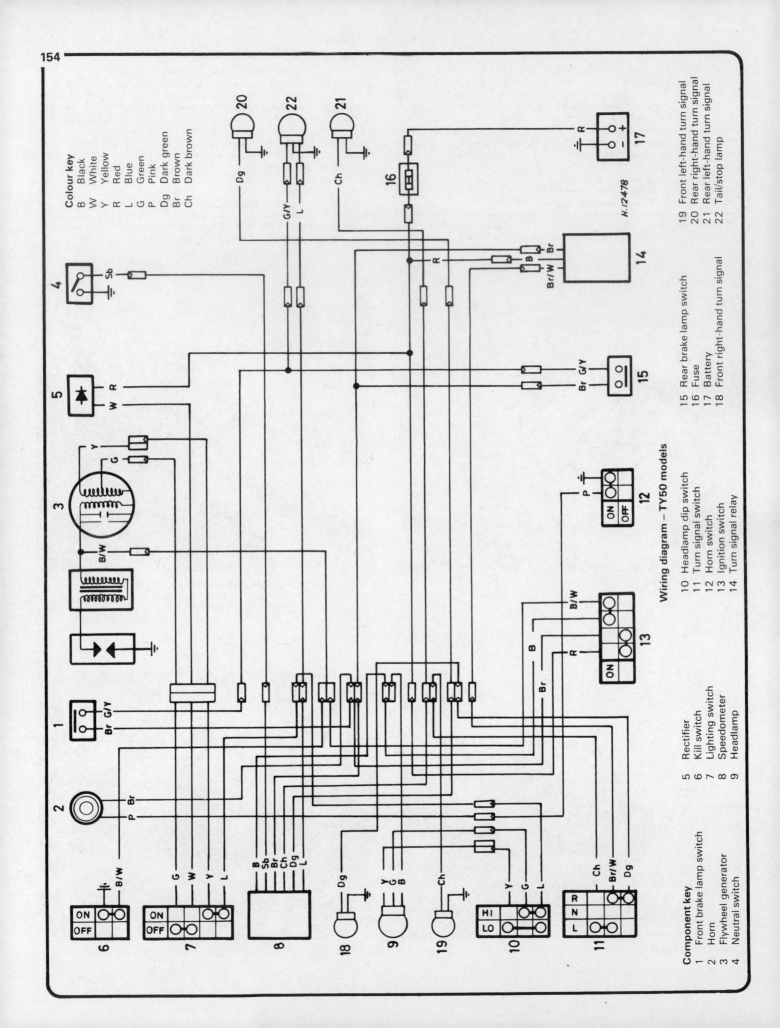

Wiring diagram – TY50 models

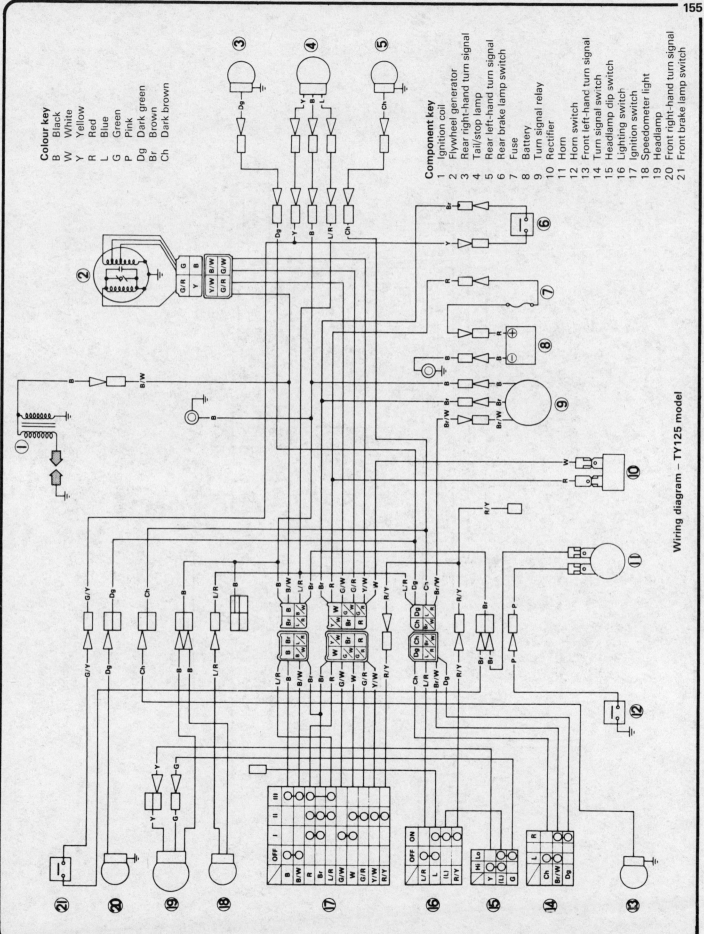

Colour key

B	Black
W	White
Y	Yellow
R	Red
L	Blue
G	Green
P	Pink
Dg	Dark green
Br	Brown
Ch	Dark brown

Component key

1 Ignition coil
2 Flywheel generator
3 Rear right-hand turn signal
4 Tail/stop lamp
5 Rear left-hand turn signal
6 Rear brake lamp switch
7 Fuse
8 Battery
9 Turn signal relay
10 Rectifier
11 Horn
12 Horn switch
13 Front left-hand turn signal
14 Turn signal switch
15 Headlamp dip switch
16 Lighting switch
17 Ignition switch
18 Speedometer light
19 Headlamp
20 Front right-hand turn signal
21 Front brake lamp switch

Wiring diagram – TY125 model

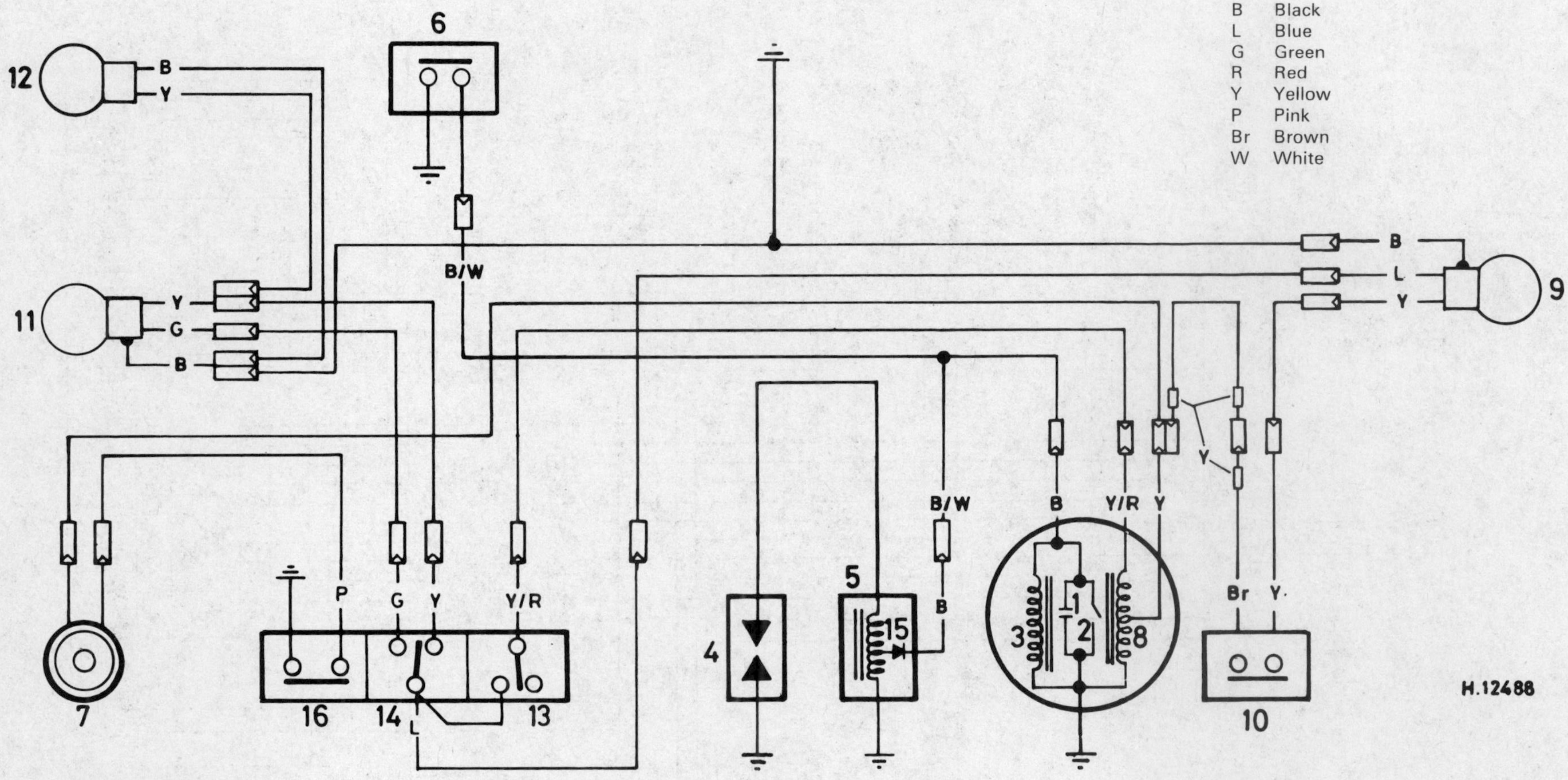

H.12488

Colour key

B	Black
L	Blue
G	Green
R	Red
Y	Yellow
P	Pink
Br	Brown
W	White

Wiring diagram – TY175 model

Component key

1	Condenser
2	Contact breaker
3	Ignition source coil
4	Spark plug
5	Ignition HT coil
6	Kill switch
7	Horn
8	Lighting source coil
9	Stop/tail lamp
10	Brake lamp switch
11	Headlamp
12	Main beam warning lamp
13	Lighting switch
14	Headlamp dip switch
15	Diode
16	Horn switch

Metric conversion tables

Inches	Decimals	Millimetres	Millimetres to Inches		Inches to Millimetres	
			mm	Inches	Inches	mm
1/64	0.015625	0.3969	0.01	0.00039	0.001	0.0254
1/32	0.03125	0.7937	0.02	0.00079	0.002	0.0508
3/64	0.046875	1.1906	0.03	0.00118	0.003	0.0762
1/16	0.0625	1.5875	0.04	0.00157	0.004	0.1016
5/64	0.078125	1.9844	0.05	0.00197	0.005	0.1270
3/32	0.09375	2.3812	0.06	0.00236	0.006	0.1524
7/64	0.109375	2.7781	0.07	0.00276	0.007	0.1778
1/8	0.125	3.1750	0.08	0.00315	0.008	0.2032
9/64	0.140625	3.5719	0.09	0.00354	0.009	0.2286
5/32	0.15625	3.9687	0.1	0.00394	0.01	0.254
11/64	0.171875	4.3656	0.2	0.00787	0.02	0.508
3/16	0.1875	4.7625	0.3	0.01181	0.03	0.762
13/64	0.203125	5.1594	0.4	0.01575	0.04	1.016
7/32	0.21875	5.5562	0.5	0.01969	0.05	1.270
15/64	0.234375	5.9531	0.6	0.02362	0.06	1.524
1/4	0.25	6.3500	0.7	0.02756	0.07	1.778
17/64	0.265625	6.7469	0.8	0.03150	0.08	2.032
9/32	0.28125	7.1437	0.9	0.03543	0.09	2.286
19/64	0.296875	7.5406	1	0.03937	0.1	2.54
5/16	0.3125	7.9375	2	0.07874	0.2	5.08
21/64	0.328125	8.3344	3	0.11811	0.3	7.62
11/32	0.34375	8.7312	4	0.15748	0.4	10.16
23/64	0.359375	9.1281	5	0.19685	0.5	12.70
3/8	0.375	9.5250	6	0.23622	0.6	15.24
25/64	0.390625	9.9219	7	0.27559	0.7	17.78
13/32	0.40625	10.3187	8	0.31496	0.8	20.32
27/64	0.421875	10.7156	9	0.35433	0.9	22.86
7/16	0.4375	11.1125	10	0.39370	1	25.4
29/64	0.453125	11.5094	11	0.43307	2	50.8
15/32	0.46875	11.9062	12	0.47244	3	76.2
31/64	0.484375	12.3031	13	0.51181	4	101.6
1/2	0.5	12.7000	14	0.55118	5	127.0
33/64	0.515625	13.0969	15	0.59055	6	152.4
17/32	0.53125	13.4937	16	0.62992	7	177.8
35/64	0.546875	13.8906	17	0.66929	8	203.2
9/16	0.5625	14.2875	18	0.70866	9	228.6
37/64	0.578125	14.6844	19	0.74803	10	254.0
19/32	0.59375	15.0812	20	0.78740	11	279.4
39/64	0.609375	15.4781	21	0.82677	12	304.8
5/8	0.625	15.8750	22	0.86614	13	330.2
41/64	0.640625	16.2719	23	0.09551	14	355.6
21/32	0.65625	16.6687	24	0.94488	15	381.0
43/64	0.671875	17.0656	25	0.98425	16	406.4
11/16	0.6875	17.4625	26	1.02362	17	431.8
45/64	0.703125	17.8594	27	1.06299	18	457.2
23/32	0.71875	18.2562	28	1.10236	19	482.6
47/64	0.734375	18.6531	29	1.14173	20	508.0
3/4	0.75	19.0500	30	1.18110	21	533.4
49/64	0.765625	19.4469	31	1.22047	22	558.8
25/32	0.78125	19.8437	32	1.25984	23	584.2
51/64	0.796875	20.2406	33	1.29921	24	609.6
13/16	0.8125	20.6375	34	1.33858	25	635.0
53/64	0.828125	21.0344	35	1.37795	26	660.4
27/32	0.84375	21.4312	36	1.41732	27	685.8
55/64	0.859375	21.8281	37	1.4567	28	711.2
7/8	0.875	22.2250	38	1.4961	29	736.6
57/64	0.890625	22.6219	39	1.5354	30	762.0
29/32	0.90625	23.0187	40	1.5748	31	787.4
59/64	0.921875	23.4156	41	1.6142	32	812.8
15/16	0.9375	23.8125	42	1.6535	33	838.2
61/64	0.953125	24.2094	43	1.6929	34	863.6
31/32	0.96875	24.6062	44	1.7323	35	889.0
63/64	0.984375	25.0031	45	1.7717	36	914.4

Conversion factors

Length (distance)
Inches (in)	X 25.4	= Millimetres (mm)	X 0.0394	= Inches (in)
Feet (ft)	X 0.305	= Metres (m)	X 3.281	= Feet (ft)
Miles	X 1.609	= Kilometres (km)	X 0.621	= Miles

Volume (capacity)
Cubic inches (cu in; in^3)	X 16.387	= Cubic centimetres (cc; cm^3)	X 0.061	= Cubic inches (cu in; in^3)
Imperial pints (Imp pt)	X 0.568	= Litres (I)	X 1.76	= Imperial pints (Imp pt)
Imperial quarts (Imp qt)	X 1.137	= Litres (I)	X 0.88	= Imperial quarts (Imp qt)
Imperial quarts (Imp qt)	X 1.201	= US quarts (US qt)	X 0.833	= Imperial quarts (Imp qt)
US quarts (US qt)	X 0.946	= Litres (I)	X 1.057	= US quarts (US qt)
Imperial gallons (Imp gal)	X 4.546	= Litres (I)	X 0.22	= Imperial gallons (Imp gal)
Imperial gallons (Imp gal)	X 1.201	= US gallons (US gal)	X 0.833	= Imperial gallons (Imp gal)
US gallons (US gal)	X 3.785	= Litres (I)	X 0.264	= US gallons (US gal)

Mass (weight)
Ounces (oz)	X 28.35	= Grams (g)	X 0.035	= Ounces (oz)
Pounds (lb)	X 0.454	= Kilograms (kg)	X 2.205	= Pounds (lb)

Force
Ounces-force (ozf; oz)	X 0.278	= Newtons (N)	X 3.6	= Ounces-force (ozf; oz)
Pounds-force (lbf; lb)	X 4.448	= Newtons (N)	X 0.225	= Pounds-force (lbf; lb)
Newtons (N)	X 0.1	= Kilograms-force (kgf; kg)	X 9.81	= Newtons (N)

Pressure
Pounds-force per square inch (psi; lbf/in^2; lb/in^2)	X 0.070	= Kilograms-force per square centimetre (kgf/cm^2; kg/cm^2)	X 14.223	= Pounds-force per square inch (psi; lbf/in^2; lb/in^2)
Pounds-force per square inch (psi; lbf/in^2; lb/in^2)	X 0.068	= Atmospheres (atm)	X 14.696	= Pounds-force per square inch (psi; lbf/in^2; lb/in^2)
Pounds-force per square inch (psi; lbf/in^2; lb/in^2)	X 0.069	= Bars	X 14.5	= Pounds-force per square inch (psi; lbf/in^2; lb/in^2)
Pounds-force per square inch (psi; lbf/in^2; lb/in^2)	X 6.895	= Kilopascals (kPa)	X 0.145	= Pounds-force per square inch (psi; lbf/in^2; lb/in^2)
Kilopascals (kPa)	X 0.01	= Kilograms-force per square centimetre (kgf/cm^2; kg/cm^2)	X 98.1	= Kilopascals (kPa)

Torque (moment of force)
Pounds-force inches (lbf in; lb in)	X 1.152	= Kilograms-force centimetre (kgf cm; kg cm)	X 0.868	= Pounds-force inches (lbf in; lb in)
Pounds-force inches (lbf in; lb in)	X 0.113	= Newton metres (Nm)	X 8.85	= Pounds-force inches (lbf in; lb in)
Pounds-force inches (lbf in; lb in)	X 0.083	= Pounds-force feet (lbf ft; lb ft)	X 12	= Pounds-force inches (lbf in; lb in)
Pounds-force feet (lbf ft; lb ft)	X 0.138	= Kilograms-force metres (kgf m; kg m)	X 7.233	= Pounds-force feet (lbf ft; lb ft)
Pounds-force feet (lbf ft; lb ft)	X 1.356	= Newton metres (Nm)	X 0.738	= Pounds-force feet (lbf ft; lb ft)
Newton metres (Nm)	X 0.102	= Kilograms-force metres (kgf m; kg m)	X 9.804	= Newton metres (Nm)

Power
Horsepower (hp)	X 745.7	= Watts (W)	X 0.0013	= Horsepower (hp)

Velocity (speed)
Miles per hour (miles/hr; mph)	X 1.609	= Kilometres per hour (km/hr; kph)	X 0.621	= Miles per hour (miles/hr; mph)

Fuel consumption*
Miles per gallon, Imperial (mpg)	X 0.354	= Kilometres per litre (km/l)	X 2.825	= Miles per gallon, Imperial (mpg)
Miles per gallon, US (mpg)	X 0.425	= Kilometres per litre (km/l)	X 2.352	= Miles per gallon, US (mpg)

Temperature
Degrees Fahrenheit = (°C x 1.8) + 32   Degrees Celsius (Degrees Centigrade; °C) = (°F - 32) x 0.56

*It is common practice to convert from miles per gallon (mpg) to litres/100 kilometres (l/100km), where mpg (Imperial) x l/100 km = 282 and mpg (US) x l/100 km = 235

Index